Louis Bauer

Lectures on Causes, Pathology, and Treatment of Joint Diseases

Delivered at the McGill University Medical College, Montreal, Canada

Louis Bauer

Lectures on Causes, Pathology, and Treatment of Joint Diseases
Delivered at the McGill University Medical College, Montreal, Canada

ISBN/EAN: 9783337190378

Printed in Europe, USA, Canada, Australia, Japan

Cover: Foto ©berggeist007 / pixelio.de

More available books at **www.hansebooks.com**

LECTURES

ON

CAUSES, PATHOLOGY, AND TREATMENT

OF

JOINT DISEASES.

DELIVERED AT THE McGILL UNIVERSITY MEDICAL COLLEGE,
MONTREAL, CANADA,

BY

LOUIS BAUER, M. D., M. R. C. S., ENG.,

Professor of Anatomy and Clinical Surgery; Licentiate of the New York State Medical Society;
Member of the New York Pathological Society; of the American Medical Association;
Corresponding Fellow of the London Medical Society; Late Health Officer
of the City of Brooklyn, etc., etc.

REPRINTED FROM THE CANADA MEDICAL JOURNAL.

NEW YORK:
WM. WOOD & CO., 61 WALKER STREET.
1868.

INTRODUCTION.

In the succeeding pages I have set forth the results of my researches into the causes, pathology, and treatment of articular diseases. Whether they are equivalent to the labours of upwards of twenty years, I cheerfully leave to the verdict of the unbiassed portion of the profession.

But those will be able to form a correct estimate of my humble efforts in this line of scientific culture, who compare the crude state of surgical knowledge on the subject then existing, with the rapid strides it has made since the ground was first broken and prepared for the new seed.

I have certainly passed through an eventful period, full of contentions against preconceived dogmas, and defended with the pertinacity of theological fanaticism; but I had also my gratifications when the new doctrines, elicited by careful pathological and clinical investigations, forced themselves with irresistible logic into legitimate acknowledgment, and now actuate the surgical practice of the most prominent standard-bearers of professional advancement both at home and abroad.

This acknowledgment is quite sufficient to satisfy literary ambition, and I can afford to treat with forbearance the literary piracy that has availed itself, without due recognition, of the results of my labours.

The opportunity of placing my literary products on this subject on record in a more coherent and complete form, I owe to the kind invitation of the Medical Faculty of the McGill University of Montreal, and I cannot allow the occasion to pass without expressing my grateful appreciation of the liberal hospitality with which I have been treated by every member thereof, and of the leading practitioners of Montreal in general.

Mr. Gross, of Montreal, and Mr. Ford, of New York, have materially assisted me with their skill in perfecting the mechanical contrivances which I employ in the treatment of joint diseases.

CONTENTS.

PAGE

INTRODUCTORY REMARKS.................................... v

LECTURE I.

CAUSATION OF JOINT DISEASES.

The strumous theory untenable.—Cruveilhier's experiments.—No pauperism, but joint diseases prevalent in the United States.—Frequency of joint diseases in childhood, among boys, in cities, and more northern latitudes.—Their rarity among adults, in the female sex, and in the south.—The anti-scrofulous treatment utterly worthless.—Local and mechanical treatment effective.—Traumatic injuries the chief cause of infantile joint diseases..................................... 2

LECTURE II.

ANATOMICAL CHARACTER OF JOINT DISEASES.

Chondritis of rare occurrence, if at all.—Structure of Synovial lining.—Its susceptibility to morbid action.—Bichet's experiments.—Periosteum.—Physiological and pathological relation to infantile joints.—Epiphyses, their peculiar maintenance, and exposure to traumatic injuries....... 14

LECTURE III.

CLINICAL CHARACTER OF JOINT DISEASES.

General symptoms.—Pain, inflammatory and reflexed.—Immobility.—Spasms—Contraction.—Malposition.—Fever.—Protracted course.—Synovitis.—Effusion.—Loss of contour.—Fluctuation.—Suppuration.—Perforation.—Hydrops articuli.—Penetrating wounds.—Periostitis and ostitis.—White swelling.—Affections of the knee-joint.—Morbus coxarius...... 22

LECTURE IV.

PROGNOSIS OF JOINT DISEASES.

Axioms in.—Prognosis better now than formerly....................... 41

CONTENTS.

LECTURE V.

TREATMENT OF JOINT DISEASES.

Method of examining.—Importance of anæsthesia for diagnostic purposes. First stage.—Absolute rest of joints the first axiom.—Means of accomplishing it.—Position of affected articulation the next.—Treatment of penetrating wounds.—Second stage.—Rest and position imperative. Paracentesis of joints beneficial and harmless.—Treatment of hydrops articuli.—Tenotomy.—Free incisions.—Treatment of morbus coxarius. Wire apparatus.—Stiffened bandages.—Portative splints and braces of Davis, Vedder, Barwell, Sayre, Andrews, and Bauer.—Their respective therapeutical value.—Treatment of affections of the knee joint.—Gutter splints.—Knee brace.—Third stage.—Exsection and amputation.—Their respective indications.................................... 42

LECTURE VI.

TREATMENT OF THE SEQUELÆ OF JOINT DISEASES.

Stiffness and its treatment.—Malposition and Anchylosis.—Gradual extension.—Tenotomy.—Brisement Forcé.—Louvcier.—Dieffenbach.—Langenbeck.—Accidents.—Rhea Barton's operation.—Brainard's plan.—Deformities of Hip and Knee-joint.—Their treatment................ 67

CASES.

I.—Hygroma Bursale Traumaticum, of eight years standing: fibrous anchylosis of left knee-joint with flexed and inverted malposition..... 86

II.—Traumatic diastasis of the lower epiphysis of left femur.—Remarkable deformity and malposition of the knee-joint.—Abnormal lateral mobility.—Total resection.—Recovery............................ 89

III.—Morbus Coxarius in its third stage.—Consecutive abscess connecting with the joint.—Complete prevention of malposition............ 92

IV.—Malposition of the right limb with more than four inches shortening, the result of now extinct hip disease........................... 95

ON
THE PATHOLOGY AND TREATMENT
OF
JOINT DISEASES.

GENTLEMEN,—In compliance with your gratifying invitation I propose to discuss some important points pertaining to articular diseases. This is possibly the only subject with which I may hope to engage so distinguished an audience.

The last ten years have been fruitful of material advancement both in the pathology and in the treatment of this class of affections, and their cultivation is still vigorously and diligently pursued. Notwithstanding all the achievements in that direction, the subject still remains in a state of transition, through the tenacity with which one portion of the profession adheres to the venerable teachings of the past, and the enthusiasm with which another portion declares itself in behalf of modern ideas. The time has certainly come when an understanding should be effected by means of unbiassed critical analysis and clinical experience. With this object I enter upon the present discourse. If, through inability, I should fail of realizing my design, I may at least hope to place the subject matter in such attractive relief as to insure your permanent interest and active participation in the settlement of the pending questions.

I.
CAUSATION OF JOINT DISEASE.

On this point, there is a decided clashing of views. By far the larger number of practitioners, the leading members of the profession among them, are of the opinion that most cases of this class are the result of constitutional disorder, of which the articular affection is but the localized symptom. To this theory the most prominent authors on surgery are committed, and it is promulgated from the professorial rostrum and at the bed-side. Time and usage have even rendered it popular with the laity. A few modern enquirers, comparatively insignificant in name and position, not only take exception to this theory of causation, but assert that articular maladies are excited exclusively by local causes, and that the constitution bears no part in the causation. They further maintain that where the constitution suffers, it suffers from the ulterior effects of the local disease.

As long as etiological views on this subject so widely diverge, there can be no uniformity of treatment; nor can a compromise be effected between views so diametrically opposite. The only way of deciding between two, of which only one can be right, is to analyse the grounds upon which they are respectively placed. I hope the venture on my part in doing so will not be deemed presumptuous, for the conflict of etiology exists, and its settlement is certainly desirable. Too much has been already conceded by the old school to warrant a proud denial; and no party can feel aggrieved when appeal is made to the decision of " stubborn facts."

Scrofulosis, rheumatism, gout, syphilis, scarlatina, pyemia, and other diseases have been enumerated as constitutional causes of joint affections. To strumous disease, however, has been assigned the first rank, inasmuch as it has been linked with the numerous and diversified cases that happen during childhood. From my own experience I have to infer that not less than ninety per cent of all articular affections occur before puberty. Inasmuch as scrofulosis is not limited to childhood, and is supposed to extend beyond puberty, a few more per cent may be added to the original proportion, making a percentage of about ninety-five. Thus the theory of constitutional causation narrows itself down to the theory of strumous causation, and with this we shall have essentially to deal.

In entering upon our investigation, gentlemen, we meet with the singular fact, that notwithstanding the general acceptance of, and acquiescence in, the stated theory, nobody seems to know accurately what strumous disease really is. There are certainly no two writers that fully agree in its definition, nor does scrofulosis rest upon any firm pathological base. Even its clinical character is rendered so indefinite that implicit faith and a goodly stretch of imagination are required to realize its attributes. This is the status of modern literature on the subject, and in extending our researches over a more remote literary period, we are not less surprised to find that the scrofulosis of the present is a materially different malady from that of the past. The pathological school of the humoralists has identified this disease with a distinct morbid principle, a *materia peccans*, contaminating nutrition throughout, and stamping all other incidental lesions with its peculiar unalterable character. The followers of that school very consistently resorted to starvation, vegetarianism, and to mercurial and antimonial preparations, for the purpose of freeing the system of that *deus ex machina*. With the physiological school the agent of strumous disease was mollified to a mere imperfect formation of proteine compounds. They very wisely adopted opposite treatment with a view to regulate the chemical transactions of the body, and to correct the catalytic combinations of the proteine. Both schools accepted perverted hygiene and diet as the remote causes of strumous disease, and consistently believed that it was a disease of pauperism. Again: both schools insisted upon strumous diathesis and an hereditary transmission. These last views are fully compatible with the humoralist principle of pathology, but indefensible from the stand-point of the physiological school. Certain appearances of patients may indicate perverted nutrition, and a morbid principle, thereby engendered, may, like syphilis, be transmitted to generations. But a diathesis for the formation of low-graded proteine combinations is a senseless construction, and the hereditary transmission of such compounds is equally without meaning and inconsistent with the chemical tenacity and restitutive powers of individual life.

Science in its advancement has already made some substantial inroads upon the strumous domain, and narrowed its borders at some vulnerable points. Porrigo capitis and sycosis menti, formerly claimed as specific strumous forms, have now been proven to be caused by insignificant vegetable parasites. The very prototype of scrofulosis, viz., keratitis scrofulosa, has been reclaimed by modern ophthalmologists as an independent and exclusive local lesion readily yielded to local appliances. And new incursions are threatened from other sides. Help was evidently needed to uphold the loose cohesion of the scrofulous architecture and to save it from pathological downfall. It was but too readily found in tuberculosis. By incorporating the latter with strumous disease, some anatomical tangibility was secured. Gradually the new pathological element has prevailed so completely, that but the name of the old scrofulous doctrine remains. In talking about strumous infiltration, *tubercular infiltration* is meant; and in fact in its former and present application, the tubercular element has completely superseded the strumous one. The transition from one to the other has been effected so clandestinely as to be noticed but by very few. The alliance between scrofulosis and tuberculosis proves, if anything, that neither had ever acquired a self-sustaining existence. Both diseases are clinically and anatomically different in character. One is said to prevail among children, the other amongst adults; and only exceptionally is this rule reversed. The organ which one chooses is but rarely sought by the other. Their very presumed causes differ most essentially,—one said to be the result of poverty and sanitary defects; the other having no respect for gradations of wealth and station. They differ even in geographical distribution. Notwithstanding all these differences, they are, by tacit understanding and acquiescence identified as the same disease. It would be unjust, however, to say that this transition has been effected totally without opposition. Of late the pathological character of tuberculosis has been subjected to various and close investigations. Its identity with pus has been asserted by Cruveilhier. The results of his experiments upon rabbits demonstrate at least this much, that pus is susceptible of undergoing the very same metamorphosis as tubercle, from the semi-fluid condition to perfect innocuous calcification. The strongest advocates of genuine tuberculosis have been forced to admit that there are often pus corpuscles, where the external appearance of the object denotes tubercular substance. Few authors have had better opportunities of studying the pathological anatomy of bone and joint diseases than Guret of Berlin, his investigations extending even over the veterinary field. If I correctly interpret his statement, he has met with no tubercle in joints and bones at all. What other authors had pronounced to be tuberculer infiltrations and caverns, he recognised as purulent infiltration the result of osteo-myelitis, and as bone abscess the sequence of circumscribed ostitis. And Virchow, one of the most esteemed pathologists of our time, considers himself justified in stating that tubercle is fully compatible with the acknowledged changes of inflammatory products. Again, gentlemen, is there any peculiarity about tuberculosis that could be established and accepted?

You are aware that the so called tubercular cell has been asserted, but the microscope has failed to prove its reality. If the microscope cannot substantiate any peculiarity, how much less can the naked eye! For there is certainly no difference in appearance between tubercular matter and cheesy pus, and the suspicion of identity must necessarily accrue from such conformity. At any rate our knowledge on the subject is not final and exhaustive; and we may justly look for further disclosures rather detrimental to, than confirmatory of, the genuine character of tuberculosis.

But, to return to the starting point of our discourse, I shall find ample occasion to show, that the stumous theory in its practical application to articular diseases, is worthless and rather injurious than otherwise, as it certainly has long diverted us from a course of investigation that alone could lead to practical results.

Consistently with the received opinions the lower classes of society must come in for their full share of joint affections simply because they are supposed to contend with poverty and hygienic neglect. If this assertion had any show of correctness, it would imply that where we find joint diseases, there we ought to expect poverty and hygienic neglect. But clinical experience in a great measure contradicts the assertion. These affections happen in all classes of society. They do not pass the mansions of the rich, nor are the agricultural districts exempt from their visitations. Yet with all it must be allowed that there is, in the abject domestic condition of the industrial classes of Europe, a plausible reason for assuming that they are more subject to chronic derangements of nutrition than the wealthy portion of society. Nor can the action of such nutritive derangements upon local diseases be altogether denied. At any rate, our pathological associations tend to confirm this supposition; though it may be clinically difficult to qualify the exact measure of those constitutional colourings of local lesions. Those who have had the opportunity of personally investigating the actual social status of the European proletariate and pauperism agree that it is deplorable in the extreme. They occupy in cities the worst of dwellings, in the lowest of quarters; their rooms are overcrowded, their articles of food are of inferior quality; multitudes subsist from offal; their opportunities for cleanliness are limited and little resorted to; their very existence is a contest for the necessaries of life. Many of the working classes and paupers domiciliate in places inaccessible to air and sunlight, in damp, and musty basements where but fungi thrive.* The combined effects of these unfavorable surroundings upon mind and body are so appalling to the humanitarian as to be remembered with painful sympathy. They give rise to the most aggravated forms of so called strumous disease with which the public hospitals and dispensaries are crowded. It is but natural to associate so conspicuous a morbific agency with a class of diseases seemingly devoid of other causes, and reacting heavily upon the nutritive standard of the patient.

---

* According to the latest statistics, 10 per cent. of the entire population of Berlin, live in cellars and basements.

In contemplating the financial condition of the same classes in the United States, we have no difficulty in finding an entirely reversed status. Here the demand for labor far exceeds the supply, and its compensation has therefore for years past been very remunerative, so as to furnish ample income to every individual who aspires to an honest living by handiwork. The "Trades Associations" have, under these circumstances readily succeeded in controlling employers and in imposing upon them their own terms for labour. However premature the eight hour labour movement may have been, this much is to be inferred from it, that the working classes are almost the sole arbiters of their own affairs, much to the oppression of the other factor of industry. So great has been the demand for hands, as to necessitate the employment of thousands of women and children. Nothing serves as better evidence of the financial thrift of labour than the acknowledged prosperous condition of the Savings Bank. Hence the domestic state of the working classes is infinitely superior to and beyond all comparison with that of their trans-Atlantic order. In fact the humblest labourer here finds himself in the possession of enjoyments which would be estimated as luxuries in Europe. However imperfect the tenement houses may be when compared with the dwellings of the wealthier classes, still they are comparatively spacious, well-lighted and accessible to current ventilation. The food of the working classes is bounteous and wholesome, and there are very few families but have animal food at least once a day. Copious water supply to tenements ensures all facilities for cleanliness; and public baths are accessible to all at a moderate rate. A glance at the attire of our industrial classes on a Sunday, gives us volumes of proof of the comparatively easy circumstances by which they are surrounded. What might have been anticipated *a priori* from their superior conditions is confirmed by practical observation, viz., that our industrial classes exhibit a better general health, a robust appearance, and none of those excessive forms of nutritive derangement which are comprised under the collective term of strumous disease. The contrast existing for instance between the populations of New York and Vienna can scarcely be overdrawn. In the Austrian metropolis almost every person one meets looks sallow, anemic, attenuated, physically impoverished, afflicted with swellings, ulcerations, and cicatrices of the cervical glands, of which in our midst there is hardly a trace.

The comparison to which I have drawn your attention, gentlemen, is between Europe and the United States, with which I am best acquainted. Whether my remarks apply equally to your prosperous Provinces, you can decide best.

Notwithstanding the superior advantages, facilities, and prosperity of our industrial classes, and notwithstanding the fact that scrofulosis in general has found amongst them but a limited ground of development, we meet, at least in the Northern States, with numerous cases of articular diseases for which constitutional causes cannot be assigned. What therefore is plausible for Europe is inadmissible with us, and this very circumstance was the first shock which unsettled my belief in the theory of strumous causation. In defence of the old theory it may be urged that tuberculosis prevails in the United States, and satisfactorily accounts for the occurrence of joint diseases. Such an argument can not be accepted as tenable, though the facts appropriated as premises may be conceded. For it so happens that tuberculosis is met with North and South, and apparently much more frequently in the latter. Among the negroes of the South, for instance, glandular affections are quite common and easily accounted for by their principal vegetable diet and hygienic indifference. If therefore the proposition be correct it will follow that joint diseases are more frequent in the South and especially amongst negroes than in the Northern section of the country. This is however not the case: on the contrary the further one proceeds South the less he meets with articular diseases; and according to the statements of competent surgeons of that region, they become perfect rarities near the Bay of Mobile, the Gulf of Mexico, and the West Indies. But irrespective of this geographical limitation of joint diseases, we have a right to demand ocular demonstration of the *tubercular deposit* alleged to be the *corpus delicti*. They are very few physicians who pretend to have seen tubercle in the affected structures. Thus, for instance, Professor Gross, who is one of the warmest advocates of the theory of tubercular causation, owns that he has never met with tubercular depositions in joints. He finds sufficient evidence for his opinion in the fact that a patient dies from tuberculosis after having suffered from joint disease. This sort of logic must pass for what it is worth. It has never converted me. For by the same reasoning we might come to the conclusion that a furunculus, a paronychia, or a fracture, happening to a consumptive patient, are of a co-ordinate character with tuberculosis of the lungs.

Gentlemen, I have submitted to your mature consideration my doubts as to the correctness of the time-honoured and prevailing opinion of strumous and tubercular causation. All I can desire of you is to look upon my arguments as suggestive. For my part I have bid adieu for ever to the old theory as an unsafe guide.

Now if the facts adduced are true, and my reasoning consistent with them, and if I have made out a clear case against the strumous or tubercular causation of joint diseases, it follows that there must be causes other than those heretofore assigned. To find them out and to prove them as such will be " the next business in order."

I have already observed that about ninety per cent of all articular affections fall upon the period of infantile development. The proportion is however very different in different ages of childhood. An articular disease is certainly a rarity among infants,—we seldom see it before the expiration of the third year. From that age upwards to the fifth year, these affections become more numerous and attain perhaps their highest numerical proportion at the sixth. Then they commence to diminish gradually, and at about the tenth year they are reduced to but few recent cases. Towards puberty these are probably as rare as during the infantile period. I need not state that these facts are based upon a careful statistical record of my own, and are borne out by the experience of well employed surgeons. I think it is apparent that the strumous theory does not offer a satisfactory explanation of these facts, for the prevalence of the disease is not supposed to be restricted to any particular period of childhood. We must therefore look for a more consistent explanation. The period of infancy is that of special parental protection. The child is mostly under direct charge of the mother or nurse, independent locomotion not having then commenced. The second and third year of infantile life enjoy less or more the same protection against accidents and injuries. With the fourth year a new epoch commences. The child is curious and inquisitive; it wishes to examine and to touch everything; it climbs upon chairs and tables; it trusts to its own guidance and escapes from the protecting eye of its mother; and it is thus exposed to all sorts of falls and mishaps. With advancing age and knowledge of its surroundings the child becomes more appreciative of danger, and more careful and timorous in its ventures. At a later period, when judgment and prudence assume their sway, accidents and particularly falls become of rarer occurrence. Reasoning from these facts I cannot but conclude to regard traumatic injuries as the sufficient cause of joint diseases during childhood.

With this supposition coincides a cordon of additional facts equally demonstrative, viz:

1. Joint diseases are not limited to any particular class of the population, nor to cities; on the contrary they occur amongst all classes of society and in agricultural districts as well as in the densely populated foci of industry.

2. Joint diseases conform to certain latitudes.

3. Certain joints are more often affected than others.

4. Boys are more subject than girls, and sanguine and impulsive children more than phlegmatic and indolent.

5. We rarely fail to trace the attack to traumatic antecedents.

6. Constitutional treatment *per se* has proved of no avail in articular affections.

7. In fine, positive results follow the exclusive local treatment of these lesions.

At 2 I do not mean to imply that climate exercises any direct or specific influence upon the numerical distribution of articular diseases, notwithstanding the undeniable facts previously adduced. But inasmuch as the temperament, usages, diet, domestic habitations, tastes, employments, &c., of the inhabitants differ according to latitude, we may be justified in speaking thus of the generative causes of disease. In comparing therefore the Northern and Southern States of the American Union we notice differences in this respect most material in their ulterior pathological consequences. The temperament of the purely Southern people is less sanguine and excitable than that of their Northern compatriots. The calmness of the Southern man is the result of his climatic constitution, and is in every respect natural, whereas the imperturbability of the New Englander is the effect of incessant social and religious discipline. The diet in one section is greatly farinaceous, in the other more nitrogenous. The habitations of the one are spacious but low, whereas the other dwells in four story buildings. There the streets and the environs of dwellings are left as nature provides; here they are paved and improved in various ways with hard surfaces. Ease has pervaded society in the South, whereas ours has been marked by constant bustle, expansion, restless and ambitious strife and collision of interests. Our employments are greatly those of a commercial and manufacturing people, theirs are those of an agricultural community. In other words our pursuits engender toil, emulation and egotism, while the ircondition is simple, calm, and primitive. The same contrast exists less or more between the inhabitants of cities and agricultural districts. What bearing, you may wonder, have these differences upon the statistics of joint affections? Simply this that a Northern child is more impulsive, ambitious, and quarrelsome, because he is confined, restricted in space, imposed upon and brought into collision with other children. His animal diet renders him stronger and more irritable. Hence his liability to casualties. Again a fall from a high staircase, or from a horse, waggon, fence, &c., to a hard side-walk, pavement or ice occasions more serious effects than the same fall upon soft ground.

At 3 it is to be noted that among all joint diseases those of the knee are most numerous; next in number come those of the hip joint; next those of the bones and joints of the spine; then those of the elbow; then those of the tibio-tarsal articulation, &c. These well known and acknowledged facts are not accidental, and the old theory fails to account for them.

It has always been alleged that strumous disease has particular affinity for the spongy and reticular structure of bones. If this be so, the tarsal, carpal, and vertebral bones should engender the disease more readily than any other portion of the skeleton. Yet as we have seen the numerical preponderance happens at the knee and hip articulations, both these joints being more than any other exposed to injury by falls, blows, and other accidents.

The proposition under the heading 4 needs no special comment. The fact that boys are more subject than girls to articular affections must be accounted for by their greater exposure to injuries. It is incompatible with the theory of strumous causation, because girls are more exposed than boys to the causes of that disease. At proposition 5 it is worthy of recollection that at certain periods of childhood accidents are of very common occurrence, though they are generally disregarded as causes of disease, unless they immediately eventuate in great pains, contusions, wounds or fractures. The proof of connection is sometimes difficult, because weeks and months may elapse before the pathological effects clearly manifest themselves. In rare cases one follows the other so closely that the mutual relation is patent and unmistakeable. That apparently slight injuries may suffice to lead to grave consequences, I have had frequent opportunities of observing. Allow me to relate but two instances in exemplification.

A little girl fell backward flat upon the sidewalk. She immediately experienced violent pain at a certain portion of the spine, and had to be carried home. I saw her soon after the fall. One of the spinous processes (the 5th dorsal) not only projected perceptibly, but was painful to the touch. The advice to keep the patient in the recumbent posture for at least three months was followed but for a short time, and the child was permitted to resume locomotion. At the end of six weeks, during which time the dorsal protrusion had noticeably increased, I was again invited to see the case. The little girl was then suffering from intense pleuritis of the left side, which eventuated within three days in copious exudation into the pleural cavity with dislodgment of the heart. Death soon ensued.

The view I held and expressed was that the recent disease was connected with the fracture of the spine; that most probably an abscess had formed at the injured point in the column, and had discharged its contents into the pleural sac. The father, in order to relieve his mind from the indirect imputation of neglect, repressed his aversion to an autopsy. I need not assure you, gentlemen, that my diagnosis was in every particular verified. There was, indeed, a fracture of the fifth dorsal vertebra, though of very limited extent, a mere chipping off of a wedge-shaped fragment still connected with the next lower intervertebral fibro-cartilage. There was next an abscess in front of the fracture and beneath the periosteum, with, as it were, two compartments, one on either side of the spine, communicating through the fracture. The left compartment, the larger of the two, had effected a perforation into the left pleural cavity. Besides this, disintegrations of bone, cartilage, and adjacent structures in general occupied the affected locality.

The other patient was a middle-aged man, a music teacher, of German extraction. When under the temporary influence of liquor, he fell from an elevation of about five feet, and struck violently the internal circumference of his right knee joint. The intense pain that set in forthwith, soon sobered him, and impressed him strongly with the apprehension of grave injury to the articulation. A physician was immediately called, but failed to discover any injury. I saw the patient the third day after the accident. There were no superficial traces left by the fall. The articulation was hot, swelled, flexed, and extremely tender to the touch. From time to time, spastic oscillations appeared, and terrified the patient, who was pale and dejected from want of food and rest. I placed him under chloroform, extended the extremity, and secured the position by appropriate appliances. The trouble yielded without any further treatment; and, for aught I know, the patient recovered from an attack that might have permanently affected the articulation.

The interval of time between cause and effect, is, after all, more apparent than real. Many cases, especially those of affections of the spine, commence in so insidious a manner, and the initiatory symptoms are so general and indefinite, as to be excusably misinterpreted not only by the parents, but even by the professional attendant. Among other cases of the kind, I remember one in particular, which had puzzled the physicians for a number of months, until a correct diagnosis was obtained.

The patient is a little boy of fine organisation, of a most impressible and active nervous system. His agility and daring even to this day are extraordinary, notwithstanding the conspicuous posterior curvature which has gradually become established. He may have been five years old, or thereabouts, when he sustained a fall from a fence six feet high, causing at the time considerable alarm to him and his parents. But no perceptible disturbance of his health immediately following, all fears were dismissed and forgotten. A few weeks after the occurrence, the patient exhibited signs of general ailment, decrease of appetite, pallor, weakness, disturbed rest, irritable temper, and indisposition to join in the frolics of his playfellows. Occasionally the pulse became accelerated, with contemporaneous thirst and increase of temperature. He complained of a transient pain in the stomach. His alvine evacuations were sluggish, badly mixed, dry, of light colour, and offensive odour. The abdomen was often distended with gas. The urine was pale, and deposited a whitish sediment. These symptoms prevailed for months without material change. The diagnosis of an "affection of the liver" was not without plausibility, inasmuch as that organ had become enlarged in all its diameters. At the end of the eighth month, frequent and painful hiccough was observed, and tenderness of the back became manifest on motion of the spine. In fine, his gait became awkward, and the movements of his body restrained and stiff. He craved for rest and support, which he obtained by placing his elbows on suitable objects, and his head upon the palms of his hands. Ten months after the accident my services were called into requisition. At this juncture it was easy enough to recognize the nature of the complaint. The marked prominence of several spinous processes at the thoraco-lumbar region of the spine rendered the diagnosis both transparent and conclusive. To the experienced practitioner, it may seem surprising that the diagnosis was not sooner accomplished, and the disease of the spine arrested by appropriate means. The entire train of symptoms pointed at a local lesion of progressive tendency: and a searching examination could scarcely have failed to reveal the locality of the affection. Nevertheless when we recollect the difficulties in the premises, the aversion of children to manual examination, the disinclination of parents to see their offspring thoroughly handled by the surgeon, and last but not least the limited field of general practitioners for fully observing and becoming conversant with these insidious cases, we will be sparing in our censure even if it should be warranted. It cannot be denied that in the case submitted, there was an uninterrupted connection between the accident and the subsequent disease. I have made the same observation in many cases that have come under my charge and have no doubt that other observers have the same experience. Nevertheless I am far from denying that joint diseases may arise from constitutional disorder likewise. But according to my clinical researches their number is proportionately insignificant. In cases of this character we find originally more than one joint affected, though the disease may eventually fix itself upon one articulation. This appertains more particularly to rheumatism, gout, and especially to pyemia. When on the other hand but one joint suffers from the beginning to the end, and the constutional symptoms supervening are in conformity with the inevitable reaction of the local process upon the general system, then it is rational to infer that the local affection is of strictly local causation.

Every candid practitioner will agree with the aphorism enunciated under 6. It is certainly a simple fact that the anti-scrofulous treatment of joint diseases has disappointed both him and his patients. My own clinical training coincides with that period in which the old etiological views held unbounded sway. They consequently regulated my action at the bedside. I followed with full confidence and scrupulous exactitude the doctrines of my distinguished preceptors Rust and Von Graefe. I coveted cases of this class, which seemed to be tacitly slighted by the more experienced members of the profession. But all my efforts were in vain. I accomplished no material change that could have been claimed as the result of devoted services. My cases took the usual course to complete obliteration of the respective joints,—malposition of the affected extremities, suppuration, caries, exhaustion and death. Nay more, I had the mortification to perceive that I could but rarely control the intense pain usually attendant upon such cases. Similar admissions have been made by other experienced practitioners, and I am led to believe that the negative results of anti-scrofulous treatment of joint diseases is now generally conceded by that portion of the profession whose opinion has any value.

In the seventh aphorism, I broadly assert without fear of contradicion that in the treatment of joint diseases, local appliances scarcely ever fail of modifying or subduing the morbid process. For the last ten years I have held these views, and practically tested them at the bed side; and I can candidly and most emphatically assure you that the results thus attained have been most satisfactory in ever particular. In but few cases have I ever had any need for constitutional remedies. Most of them yielded readily to local means; and with the local improvement the prevailing constitutional disturbances subsided. When thus rest and appetite were insured, the patients increased in *weight*, and rapidly improved in appearance and feeling. I need hardly state that my therapeutic views on this point were slighted for a number of years by those men to whom the profession look up for precept and example. But when Dr. Davis' portative extension apparatus became generally known, the professional mind underwent a material change and then turned its attention to the subject. A few years ago the New York Academy of Medicine discussed the subject of hip disease at successive meetings. Most of those who participated in the discussion admitted in emphatic terms the therapeutic efficacy of that instrument, retaining at the same time the old tubercular theory of causation. Nobody seemed to notice the contradiction between theory and practice, and it was then and there that my views gained the ascendency. I simply stated on that occasion that but one could be right. "If hip disease were the consequence of strumous invasion, a portative extension of but few pounds could have no effect whatever in relieving or curing that complaint; and if it actually had the effect alleged, it would be the most undeniable proof against the constitutional character of the disease." The attempt to refute my logic was as feeble as it was unsuccessful, and from that date it may be said that the new theory was admitted to scientific citizenship. I shall not on this occasion enter more extensively upon the subject, inasmuch as I have to recur to it when speaking on the treatment of articular diseases.

II.

ANATOMICAL CHARACTER OF JOINT DISEASES

GENTLEMEN,—All the anatomical components of a joint may separately and collectively become diseased. Their morbid susceptibility varies however in a material degree. The articular cartilage occupies obviously the lowest point in the scale. In conformity with its purely physical office, it is elastic, only indifferently organized, and devoid of nerves and vessels. Its nutrition is therefore of a low order, accomplished chiefly by transudation and imbibition. Reasoning from these premises it might *a priori* be assumed that this structure possesses but a trifling susceptibility to independent morbid action. This supposition receives additional strength from experiments upon animals by Redfern, O. Weber, and others who found that neither physical violence nor chemical irritants have much lasting effect upon articular cartilage. The intervertebral fibro-cartilages are of higher organization, and are therefore endowed with a more decided susceptibility to morbid changes than those of joints. I have made clinical observations to this effect, and I have recorded one case of inflammatory disintegration of so striking a character, that no reasonable doubt could be raised against it. In advanced diseases of joints and of the spine it is impossible to determine whether the cartilage or some other structure has been first affected. The destruction is commonly so general as to leave no room for speculation. I am inclined to believe that the cartilage suffers but rarely from primary lesion, but that it often participates in the affection of the subjacent bone, and is subject to disintegration from purulent maceration.

That the cartilage displays but a passive character in the so called *arthritis deformans progressiva* is now well understood.

The synovial lining is a sort of intermediate structure. It does not conform to serous membranes with which it has heretofore been classed. Its greater thickness, albuminous secretion and layered epithelium bring it nearer to the anatomical structures of mucous membranes from which it differs by the absence of mucous follicles. The Haversian glands are no glands at all, but synovial insaculations filled with fat. Gosselin's fimbriae have thus far not met with general acceptance, nor have their functions been fully ascertained.

According to Richet the healthy synovial membrane is very vulnerable. Injections of irritating fluids into the joints of animals are promptly followed by great vascularity, hyperemia, pinkish and purple discloration, and opacity of the synovial lining with serous infiltration of the adjacent connective tissue. The vessels frequently cluster around the articular cartilage, and by anastomosis form as it were a continuous wreath from which returning twigs branch over the margin. Occasionally the synovial membrane becomes so oedematous and pouched as to circumvallate the cartilage as chemosis does the cornea. By degrees the entire surface of the joint becomes roughened and granulated. The epithelium luxuriates and is converted into pus corpuscles which are successively thrown off and the articular cavity is filled with purulent fluid (pyarthrosis); similar pathological changes may often be observed to follow penetrating wounds, with this difference however that in the beginning the synovial fluid forms a material constituent item of the discharge, and reappears occasionally when the process is subsiding. From these experiments it would seem that the synovial lining, notwithstanding its destitution of nerves and vessels, is highly susceptible to morbid action of the peracute type. But clinical experience has collected many facts to the contrary. Thus, for instance, some penetrating wounds close by first intention without inconvenience to the injured joint, although blood may have been left behind and air may have entered. Many a time have I performed articular puncture by trochar and knife without a single bad effect, having of course, as much as possible, prevented the entrance of air.

In hydrarthrosis, Nelaton has freely resorted to injection of iodine, and others have followed his example. According to their statements, only a moderate reaction usually ensues. Free incisions into affected joints have been made, checking the disease, and saving extremities. Amputations in contiguity leave always a portion of the joint, and some surgeons prefer these operations on account of better statistical returns. These facts constitute a formidable offset to the rule based upon Richet's investigations. It is not unlikely that chemical irritants, applied to a healthy articular surface, will readily lead to a rapidly advancing synovitis, and repeated applications of this sort will bring about those progressive changes, of which Richet gives so graphic an account. But it does not follow that atmospheric air would give rise to the same disturbances. According to my experience, the dangers of penetrating wounds have been altogether overrated. In the course of the last few years I have attended a considerable number of cases, many of them formidable, and have in every instance obtained satisfactory results. This may have been due, in part, to the healthy condition and tolerably good surroundings of my patients, but not less to the more appropriate treatment that has found its way into surgery. From clinical observation, however, I have received the impression that the synovial membrane has a dangerous affinity for disturbing causes of a constitutional character. Rheumatism, syphilis, and pyemia, in particular, select this structure in preference to the other components of joints. Of late much has been said and written about tubercular synovitis; Foerster has never met it, and he is certainly no superficial observer. Nor have I had an opportunity of examining a single case of this description, although I may say, without boasting, that I examine as many cases of joint diseases as any well-employed surgeon. If, moreover, tubercular synovitis is of a nature similar to that of tubercular meningitis, it means little more than initiatory changes in the subsynovial tissue towards suppuration,—namely, hyperplasy of connective tissue. Still I do not pretend to express a conclusive opinion upon what has so sedulously evaded my most inquisitive pursuit.

Some authors believe that the synovial lining suffers most severely from incidental traumatic injuries. I beg to dissent from this opinion. If both constitutional and local causes expend their force upon the synovial membrane, all joint diseases would resolve themselves into synovitis, and the other components would pass clear of primary disease. Both clinical and anatomical observation refute views so untenable. Most injuries befall the prominent portions of joints—the bones and their periosteal coverings, because they are most exposed, and because they offer static resistance. And even if the synovial sac comes in for its lesser share, the consequences cannot be beyond speedy redress. Inflammation, excited by a transient cause, would soon terminate in copious secretion of synovial fluid; and this, in turn, would be absorbed. A moderate admixture of purulent elements would not materially affect final resolution. Permanent disintegration of the synovial lining, or of the other constituents of the joint, could not well be ascribed to a comparatively trifling and transient cause.

In the anatomical consideration of joint diseases, there has not yet been assigned to the periosteum that importance which it so fully deserves. In the first place, the periosteum continues as part of the joint from one bone to the other, constituting the so-called fibrous capsule. Next, it partly covers the epiphyses and condyles of the cylindrical bones, and constitutes the means of their maintenance, growth, and development. From the first anatomical relation results the direct transmission of disease; and upon the other depends the structural condition of an essential articular component.

In the course of my surgical practice, I have observed cases of joint disease that could be traced to no other cause than traumatic periostitis. Some of them involved both limb and life. I will relate one in striking exemplification. A lad of thirteen years, in perfect health, and without any noticeable morbid diathesis, was struck with a medium-sized cobblestone at the middle of the tibial crest. Judging from the lesser age of the boy who aimed the blow, from a distance of about twelve yards, the force could not have been very considerable. The impression upon the leg was apparently insignificant. The pain was trifling, and no bruise or indentation appearing, the patient paid no attention to the injury during the succeeding five or six days, and continued at his duty as an errand-boy. Subsequently he found locomotion impracticable, his leg having become painful and so swollen that he could not get his boot on. A physician was now sent by the father of the offender. The attendant failed to penetrate the nature of the lesion. Thus twelve days more were irretrievably lost in paltry applications. When better advice was finally obtained, the disease had made considerable advance, demanding more than anything else extensive and deep incisions. These were not resorted to to a sufficient extent. I was called in at about the sixth week after the accident, and found the patient in a most critical situation, and fearfully reduced. Then no alternative to amputation remained, for the limb and the corresponding knee-joint were so extensively and irrecoverably diseased that no attempt at saving the limb could be entertained. The specimen revealed the following state:—Almost entire destruction of periosteum of the tibia, exposure and discoloration of that bone; the remaining portion of the periosteum towards the knee-joint undermined, allowing the passage of a stout probe into the articular cavity at the lower insertion of the fibrous capsule. The latter was itself perforated by ulceration at the external and posterior walls, and the joint exhibited the pathological changes of advanced pyarthrosis. The patient had a speedy recovery, and has for the last six years enjoyed the most unqualified health. Now, gentlemen, this case proves indeed more than I have claimed. Here a lad in perfect health receives an injury at a point remote from the knee-joint, which lights up an inflammation of the periosteum. Not being recognised and controlled, the inflammation proceeded to suppuration; the matter spread below the periosteum in every direction, until it reaches the capsular apparatus, and finds access to the joint. As soon as the diseased structures are removed, the patient regains his former health and strength, precluding every suspicion whatever of constitutional disease. This is certainly a clear case of traumatic periostitis, involving an articulation; and the chain of evidence is continuous from the very starting point to the finale. This case is by no means as isolated and exceptional as might be supposed, although in others the clinical history may not always be found so plain and transparent.

The foregoing belongs to a class of cases that are generally insidious and protracted. For a long time they cause but little inconvenience to the patient, and therefore they are slighted at the time when appropriate treatment could scarcely fail to arrest their progress. Thus with very little change they pass on for many months, until an acute period is reached and the joint is found to be extensively diseased. The original traumatic cause is forgotten; it appeared at most to be insignificant, and in the estimation of all parties concerned, could not have given occasion to consequences so severe. Meanwhile the constitution of the patient has materially suffered, the vital forces are depressed, the appetite has become indifferent, weight has decreased, in fact nutrition has gradually and proportionally declined, as the local disease has extended its sway. This is the history of most cases occurring during childhood, and it is this class that has been set down as the result of strumous causation, in default of any other known cause.

Now, gentlemen, *must* there not be a *general predisposition* attached to *the physical condition of infantile development*, that favors diseases of joints, and disappears at puberty? No one seems to have paid much attention to this query, and hence the preponderance of joint affections in childhood has remained unaccounted for, up to this very day. It is still an enigma unsolved.

Laying aside all the fetters of established doctrines, let us try to find out some of the anatomical differences existing between the joints of children, and those of adults. Perhaps they may furnish us the key to a correct understanding of the matter. All we meet is the epiphysal contrivance which serves wise purposes in the growth and development of the osseous architecture, but allows the epiphyses themselves to be liable to mechanical derangement. We need but to look at a vertebra composed as it is of seven different pieces held together by cartilaginous discs and periosteum. By this arrangement it is rendered a very elastic body capable of accommodating itself to many exigencies. But its resistance is limited to its elasticity, and the single pieces may under certain circumstances become disjointed or somewhat altered in mutual relation. Diastasis is a solution of continuity solely appertaining to the period of childhood.

At an early stage of infantile life the different epiphyses of the skeleton present a marked peculiarity in the mode of their maintenance, and there is reason to believe that this mode partially continues to within a short time before puberty. Careful injection of the nutrient vessels of the bones of infants and children, demonstrate pretty clearly that the epiphysis receives no vascular complement from that source. In fact the vessels pass only to, and not through the epiphysal cartilage. On the other hand the vessels that enter the epiphysis have no communication with the nutrient artery of the shaft. They are, as it seems, completely isolated from each other by the cartilaginous disc. Most epiphyses are supplied with blood from the periosteum, with which they are in part covered. Those epiphyses to which the periosteum can not approximate closely enough, have a special source of nutrition. Thus for instance the head of the femur receives its supply from a branch of the obturator artery which enters the notch of the acetabulum and accompanies the so called ligamentum teres, to its destination. The nerve takes the same course. A rather complex mode exists at the knee joint through both periosteum and the ligamenta cruciata. After the skeleton has attained its full development, and the epiphyses have become continuous with their respective bones, nutrition is perfected by anastomosis of the several vessels. But the intermediate parts of some bones seem never to achieve a full share in nutrition, thus we know that fracture of the femoral neck but rarely heals by bony union. It is very necessary that we become fully acquainted with all these physiological facts as they serve to throw light upon a field hitherto obscure.

The epiphyses constitute the most prominent part of the joints, and receive most of the violence of traumatic injuries, the soft parts being thus in a measure protected. At the limited space of contact with the offending force, the integuments and the periosteum are contused and ecchymosed, and the nerves of the joint less or more injured. The integuments may soon recover; at any rate their structural derangement would be of but little consequence. Not so with the periosteum. If the extravasation of blood takes place in the usual way, that is to say beneath the latter, it constitutes in my estimation a serious trouble. Irrespective of ecchymosis, the eventual cause of subperiosteal suppuration, the very presence of blood denotes disruption of the vessels intended to supply the nutritive demand of the epiphysis. The extent of the part borne by injuries of articular nerves (sensitive and trophic) in exciting articular diseases has as yet not been clearly ascertained. A case previously detailed gives strong evidence to this effect. The same injury to any other part of the bone might be comparatively harmless, and would generally eventuate in exfoliation, because the nutrition of the bone depends only in part on the periosteum. It would seem therefore that even apparently trifling contusions at the epiphysis should be viewed with deference and treated with becoming care. But if they give rise to subperiosteal suppuration, there is in two ways imminent danger for the joint:—first, by the matter spreading below the periosteum and forcing its way into the articular cavity; and secondly, by instituting necrobiosis of the epiphysis in part or *in toto*. The latter mode is obviously the more frequent. The destruction or detachment of the entire epiphysis by this process is very rare,—more frequently, one of the condyles is implicated, enlarged, osteoporotic, and very tender. From thence the disease radiates to the remaining structures, and thus the joint becomes compromised. I have but lately exhibited to the New York Pathological Society, a specimen illustrating this process. A small sequestrum in the internal condyle of the femur was evidently the proximate cause of the extensive trouble to the joint, amounting to an almost complete obliteration of its cavity by adhesive synovitis.

Primary diseases of the epiphysis are not of frequent occurrence, and least of all osteomyelitis.

The process of gradual destruction is most simplified at the hip-joint, and its varied phases may best be studied there. A few anatomical remarks will be necessary. The ligamentum teres must be accepted as a ligament in an anatomical point of view, on account of its being endowed with a considerable complement of fibrous structure. Besides this, however, areolar tissue and fat enter largely into its composition, encompassing the nerves and vessels passing to, and from the head of the femur. No anatomist has as yet been able to demonstrate the office of the round ligament. The head of the femur fits so accurately in the acetabulum that it is held there by atmospheric pressure, or, as others think, by cohesion. This bone may dislocate in any direction without the ligamentum teres being ruptured; it consequently places no restraint upon the movements of the thigh bone. Some instances are known where the joints lacked it altogether, without marked impediments resulting. Again it has been ruptured in the act of violent dislocation and the returned head of the thigh bone moved almost to the same perfection as before. Thus it would appear that this ligament bears no part in the action of the hip joint. Another office must have been assigned to it. To all appearance it acts as the protector of those nerves and vessels which form the nutritive apparatus of the head of the femur. Without this protection the nutrition of the femoral epiphysis could not be effected. Collectively I look upon the ligamentum teres therefore as the essential nutritive appendix of the head, and its destruction during the epiphysal period as tantamount to the destruction of the head itself. From the composition of the round ligament a high degree of susceptibility must be inferred. In fact, none of the articular components can bear any comparison to it in this respect. Besides the ligamentum teres is subject to contusion from violence to the great trochanter, whilst the thigh is in the position of adduction and eversion. And upon the trochanter falls are generally received. Boyer has already expressed the belief that morbus coxarius emanates from the round ligament; but for want of pathological facts, he did not succeed in convincing his contemporaries. The scrofulous theory very soon preponderating, overawed his views, which well deserved consideration. Perhaps no articulation has suffered more from the dogmatism of the humoralist school than the hip joint; and the fiction culminated into a system in morbus coxarius. There were explanations in it for every single symptom. Very few of these are destined to survive the present century.

It cannot be denied that morbus coxarius may possibly be caused by primary synovitis or periostitis with subsequent centripetal perforations. But the majority of cases must necessarily result from primary disintegration of the round ligament. Among the reasons for this opinion, of which I have already enumerated a few, stands in the boldest relief the pathological fact that the round ligament is invariably destroyed at a time when the remaining components of the joint have suffered but moderate disintegration. Next comes the striking fact that the head of the femur is invariably reduced excentrically in size, and in a few exceptional instances thrown off *in toto*. That the origination and frequency of morbus coxarius in childhood has the closest connection with the epiphysal construction admits of no doubt in my mind; and it explains satisfactorily the comparative rarity of this affection during adult life when the epiphysis is completely united with the shaft, its nutrition thereby perfected, and the liability to accident lessened.

Gentlemen, I shall here close my discourse on the pathology of joint diseases, and not inflict upon you a reiteration of all that is said better in the works of Sir Benjamin Brodie, Robitansky, Paget, Gurlt, and other distinguished pathologists. Moreover, the practical benefit of being thoroughly versed in the ulterior structural changes attending joint diseases, is indeed of questionable value. If you see one joint in the last stage of its malady, you have seen them all, so little difference between them is presented. My chief object has been to acquaint you with the initiatory changes of joint diseases, and thus lead you in a practical direction for the prevention of their destructive advancement. But even in this, I have had to consult brevity and terseness in order best to utilize the limited time at my disposal.

III.

CLINICAL CHARACTER OF JOINT DISEASES.

All joint diseases have some symptoms in common. Of these pain is the most prominent; usually the first to appear, and the last to disappear. Clinical observation discerns two kinds of pain—one emanating directly from the diseased structure; the other proceeding in a circuitous manner from the spinal cord, and manifesting itself in parts not directly connected with the affected articulation.

The former is known by the term of *structural* or *inflammatory* pain; the latter as *reflex*. The structural pain varies in extent, intensity, and duration, according to the tissues implicated, and to the nature and extent of the malady. In some instances the pain may occupy but a small and circumscribed place; in others it may be diffused over the entire articulation, and extend even beyond it.

Its intensity may vary from the sensation of heat and soreness, to the degree of burning, lancinating and pulsating; and be equally variable in its continuance.

The morbid condition of the affected structures does not always furnish a satisfactory explanation of the degree of pain; but too often one is out of keeping with the other. Thus, for instance, a mere ephemeral rheumatic synovitis, and in hysteric affections, the pain, for the time being, is very intense and largely diffused, whereas, in hydrarthrosis but little inconvenience to the patient arises from a similar source. The general affection of an entire articulation, with advanced disintegration of the various tissues, may exist for months, and yet be attended with comparatively little suffering, whilst on the other hand, affections apparently trifling, may create a storm of symptoms and intense agony.

In structural pain therefore, but a conditional semiotic importance can be attached. In this respect the same axiom rules as in the healing art generally—"that but the congruity of symptoms is the base of diagnosis."

Notwithstanding all this, some general rules can be recognised as a guide at the bedside.

1st. The structural pain is commonly proportionate to the nervous endowment of the tissue affected.

2nd. The pain increases and diminishes in proportion to the progress and regress of the disease.

3rd. The pain is rendered more intense by false position of the articulation.

4th. The pain increases when the affected structures become subject to centrifugal distension by effusion of whatever composition, and to irritation by pus, loose sequestra, and foreign bodies.

5th. The pain is augmented by touch and motion.

6th. Whatever induces and increases pain, hastens the advance of the articular disease, and vice versa.

The so called reflex pain is obviously of a neuralgic character. Being excited by the local disturbance, the morbid impression is conveyed to the spinal cord, the common centre of irradiation; thence it is reflected backward to the muscles appertaining to the affected joint, and sometimes to the next articulation; as for example, the almost pathognomonic pain at the knee in coxalgia.

The latter mode is rather an exception, and an isolated clinical fact, which may be explained in this manner: " that the same nerve (obturator) supplies both joints with sensitive fibres, warranting the supposition of irradiating in the closest proximity."

From the fact that the reflex pain occurs commonly during night and the sleep of the patient, it must be inferred that the trophic or ganglionic province is principally, if not exclusively involved. But a few exceptions have come to my notice to which I shall refer in due course. You are perhaps aware that I was the first observer of these reflex pains; at all events, I was the first who called attention to them, and explained their character and operation. Perhaps they might have escaped my observation as well, had I not for a time shared the same roof with patients of this class, and had not thus an opportunity been afforded me for studying this singular symptom in all its bearings.

One night, after having left my patients profoundly asleep with the lights lowered, my attention was suddenly attracted by a peculiar shriek emanating from the sick room. Within half an hour the shriek was twice repeated.

Though well acquainted with the different voices of my little patients, I could not discern to whom the cry belonged. It was in so peculiar a note, high, shrieking and short, commencing with a full intonation, and terminating as abruptly. In entering the room, I found everything and everyone as quiet as I had left them shortly before. The only noticeable change was an acceleration in the breathing of one of the patients.

Whilst thus contemplating and watching him, he again uttered the same shriek, rose into a sitting posture, rubbed his eyes, stared around with a terrified expression, and sunk back upon his bed, continuing his scarcely interrupted sleep. In another ten minutes this scene was re-enacted, with almost the same concomitants. During several of these paroxysms I observed a peculiar quiver of both the adductor and flexor muscles of the thigh. The rest of the joint was evidently disturbed by it, and the pain accompanying the quiver must have been of an agonizing character, for the patient automatically grasped the affected limb, as if to arrest the involuntary movement. His rest for the balance of the night was disturbed by moanings, and repeated attempts to changing his position. I found the aspect of the patient much changed on the following morning; he looked pallid, haggard, and prostrate; he was of morose and irritable temper, his pulse excited, and his appetite indifferent. The tenderness of his joint had signally increased. Whilst the abduction was more difficult and painful than before, the entire group of the adductor muscles was as tense as if possessed of tonic spasm.

In continuing my observations for successive years, I have seen this very symptom in almost every aggravated case of joint disease in structural affections of the spine, and in acute periostitis in the proximity of joints. In all these cases it is invariably of the same type, though varying in intensity. The greatest violence of reflex pains we observe in morbus coxarius, and in affections of the knee joint.

It is rather remarkable that the patients thus afflicted do not remember these nocturnal pains, and that the shrieks of different patients are almost invariably of the same note and duration.

It may well be said these shrieks are as characteristic of joint disease and as important in its diagnosis, as the peculiar croup tone in diphtheritic laryngitis, and the cries of a parturient woman in the last period of confinement.

As already remarked, these reflex pains occur almost exclusively during the night, and whilst the patient is dormant.

In a few exceptional cases, however, I have met the symptom under inverse circumstances. In one case (Schindler) the pains continued for several days and nights, and kept the affected member with but short intermissions, in a constant state of clonic spasms, and until the flexors of the leg had been divided.

They may be met with, irrespective of time, when contracted muscles are put upon the stretch.

Whenever the reflex pains prevail, the patient suffers most severely; loses flesh and appetite; becomes anæmic, and prostrate, and the disease of the joint progresses with marked rapidity.

According to my clinical experience, the reflex pains chiefly accompany bone diseases, and in these they are most severe. In synovitis they are certainly much milder, if at all present.

In some instances the reflex pains assume the character of genuine neuralgia, and follow the course of the principal nerves; in others they discharge their violence upon certain groups of muscles, painfully oscillating and cramping them, leaving them in a state of cataleptic tension.

With the symptom of reflex pain, two others are very soon ushered in:—

1st. *Attenuation of the affected member.*
2nd. *Muscular contraction.*

The wasting of the affected extremity is as common a symptom of articular diseases as it is conspicuous. The adipose tissue becomes rapidly diminished, and finally extinct; the muscles lose their bulk and normal contour, the bones lose in circumference and length; the extremity assumes a cylindriform shape; its growth is arrested; the animal heat is below the standard of the body, and in cold weather the extremity presents that mottled appearance which is so common in paralysis.

The symptom of attenuation is co-ordinate with that of muscular contraction, and never observed without the latter.

Among the many hypotheses advanced in explanation of this symptom, that of Barwell is about the most superficial, ascribing it to the permanent compression of the capillaries within the muscular structures. At best this theory would apply to the waste of muscles, but leaves the other structures of the extremity out of account.

Without entering into a digest of the various opinions, I shall content myself with offering my own. It requires, indeed, no great pathological acumen or diagnostic sagacity to reduce that symptom to its proper source. It consists not only in the diminution of substance, but the arrest of growth is so prominent, that impeded innervation and impeded nutrition must be charged with the mischief, for which pathology furnishes ample analogy.

In club-foot, for instance, the very same conditions prevail, the same attenuation—the same arrest of growth and development—the same reduction of temperature, co-existing with muscular contraction and malposition.

The muscular shortening in joint diseases is well known to careful observers, but its pathological character has as yet not been fully appreciated by the profession. In carefully analysing the facts in the premises, I shall encounter no difficulty in establishing views fully consistent with the nature of the symptom in question.

1st. I have already adverted to the influence of the reflex pain upon certain muscles appertaining to the affected articulation, setting them into a most agonising quiver. This symptom is, indeed, so common, that its peculiarities may be ascertained beyond a shadow of doubt.

2nd. When these muscular spasms subside, they leave its structure in a state of rigor, or stationary retraction and tenderness, which, however, gradually disappear, if no new spasms set in.

3rd. Every attempt at elongating the so retracted muscle, by gradual extension, is very painful, and not rarely it is resisted by returning spasms.

4th. Faradayism renders the state of so retracted muscles still more tender, and not seldom gives rise to greater and painful shortenings of the muscular belly.

5th. During anæsthesia the muscular retraction relaxes and allows full extension, which, in some instances, may be successfully perpetuated by appropriate appliances. In others, the retraction reappears with the cessation of the anæsthetic effect; the muscle remains tender and jerking. If, under these circumstances, the extension be persisted in, the articular disease becomes aggravated.

6th. Persistent retraction terminates in structural changes of the muscle, and destroys its expansibility, both physiologically and experimentally. Faradayism produces scarcely any excitation whatever, and chloroform anasthæsia exercises no marked influence upon its tension. Thus the muscle, having attained its maximum of contraction, and that contraction being rendered permanent by organic changes of its structure, the term contracture has been fitly applied to that condition.

Dr. Benedict, of Vienna, maintains that a constant galvanic current possesses the power not only to reduce the contraction, but to establish the physiological expansibility of muscles so affected. I have, however, not seen a single case at his clinic in the general hospital of that city that could be accepted in proof of his views.

Nor can the successful *brisement forcé*, without myotomy, pass as evidence, since the violence generally employed is quite sufficient to tear asunder all resisting structures—myolemma or muscular fibres—thus virtually accomplishing the same results as would be produced by dividing the contracted muscle.

7th. The subcutaneous division of the contracted muscle overcomes both resistance, spasm, and attending pains.

8th. The division of contracted muscles exercises the most beneficial influence upon the affected extremity, in promoting its nutrition, growth, and development. Even the muscles themselves become more bulky and susceptible to the action of Faradayism.

The contractures of muscles, force of course, the affected extremity into a position corresponding to their respective traction, and they become therefore the source of malpositions.

In all joint diseases some muscles, or group of muscles are invariably contracted to the exclusion of others. Thus for instance, in morbus coxarius, we find the adductor muscles of the thigh, and some of the flexor muscles materially shortened. Among the adductors, the pectineus; and among the flexors, the tensor vaginæ femoris, are the most implicated. In consequence of these contractions, the affected extremity is unduly flexed, and adducted and rendered apparently shorter than its fellow, the disparity being increased by the elevation and rotation of the corresponding side of the pelvis. In affections of the knee joint the biceps muscle is commonly the only one contracted, and but exceptionally the remaining flexors become involved. Hence the affected member is more or less flexed at the knee joint, and in the higher degree of flexion, the leg is rotated on its longitudinal axis, and the toes everted. This position implies an anatomical derangement of the respective parts of the joint, the external condyle of the tibia receding, and the internal, protruding in front of the joint. In affections of the tibio-tarsal and tarsal articulations, the peronœi muscles are retracted, and thereby the foot rotated so as to give it the position of talipes valgus. In affections of the wrist joint we meet with contractions of the flexor radialis and ulnaris, with abnormal flexion of the hand; sometimes but one of those muscles is shortened, and the hand has a corresponding leaning in its direction. In affections of the elbow joint the biceps muscle and the pronator teres are involved keeping the forearm in a state of pronation and flexion. In affections of the shoulder joint we notice the contraction of the pectoralis major, with adduction of the arm to the body, &c.

It is self evident that the contraction of certain muscles in certain joint diseases is by no means accidental but governed by the supply of co-ordinate nervous fibres. Schwan by his very careful and minute dissections, has fully established the fact, that such a co-ordination of nerves exists, supplying joints and muscles. And Hilton, another reliable anatomist, has affirmed that anatomical arrangement. But even without these anatomical facts, clinical observation would be justified in such an inference.

In most joint diseases there is more or less immobility. To a certain extent the immobility is of a voluntary character employed by the patient to obviate the pain caused by the exercise of the affected joint. Frequently, and in advanced cases, the immobility may arise from hydraulic pressure upon the articulating surfaces, by effusion into the joint, as may be seen in the second stage of hip disease, and in some affections of the knee joint with unyielding and thickened walls.

The deposits of osseous material around the joint, and osteophytes, will produce the same effect. Muscular contractions are a material impediment to the mobility of affected joints.

I have already referred to malposition of the respective affected articulations, as one of the general symptoms attending articular diseases, and adduced its most prominent cause. There are however other causes which occasionally bring about that result. One of them is the gradual disintegration of the epiphysis. Next the separation of the epiphysis and its dislodgement from the shaft. Another, the fracture of the epiphysis eventuating in joint disease. The last though not least is effusion within the articular cavity. The experimental injections into joints made by Weber and Bonnet demonstrate that liquids forcibly thrown into the articular cavities through an aperture of a stationary bone will force the movable part of the joint into certain positions denoting the greatest capacity of the articulation.

Similar changes in the position of joints are produced in the living body by effusions.* But in order to accomplish this the walls of the articulation require to have been rendered unyielding to the process of inflammation, in which case the effusion acts like a wedge driven between the articular surfaces. As long as the walls remain flaccid, or retain their

---

* Collateral with more or less perfect immobility.

healthy elasticity; an immense quantity may be accumulated in the joint without any effect upon its position, as is the case in ordinary hydrathrosis.

Last, I have to mention fever, as one of the common symptoms of joint diseases. This symptom is merely of temporary duration, and accompanies only the higher grades of these affections, their inflammatory periods, or at times when a mighty local irritation exists, be this through foreign bodies, sacculated pus, or the like. It generally subsides with the removal or alleviation of the local disturbance. In all these instances the fever is strictly symptomatic. Rheumatic affections of joints are however, ushered in with marked febrile excitement, which seems to form an essential part of the morbid process.

Profuse and continuous suppuration of joints is mostly attended by hectic fever, which presents the usual characteristics. But rarely do we meet with pyæmia, caused by affection of the joints. I do not think that I have seen more than a dozen cases, in all in my practice. The latest refers to a little girl, eleven years old, of very delicate constitution. From causes unknown, she was attacked almost simultaneously with an affection of the left tibio-tarsal joint, and periostitis of the corresponding tibia, both disorders eventuating rapidly in suppuration. A few weeks after the first attack, a large abscess had formed during one night at the left hip; another soon afterwards made its appearance below the right clavicle, soon to be followed by a third in the right hip.

It is yet doubtful in my mind whether this case does not come under the head of spontaneous pyæmia, a form which is seriously doubted by some authors, or whether pyæmia resulted from the original affection.

The division of joint diseases into acute and chronic forms, is rather inappropriate, because artificial. It is apt to confound the character of the affection, and has no practical value in any respect. Whether the duration of the malady, or the violence of the symptoms is the principle of division we shall find neither to be tenable.

Almost every joint disease assumes *a protracted course*, and is thus essentially *chronic*. But few exceptions can be adduced to this rule. Rheumatic synovitis may be of short duration, and characterized by violent symptoms, but joints thus affected will require months to recover their normal status. On the other hand, we observe periods of acuity, in the most chronic and protracted joint diseases, which may challenge the most acute forms known.

I suggest, therefore to drop a clinical dogmatism, worthless to the experienced surgeon, and confusing to the novice.

The symptoms by which *synovitis* is characterized, materially vary, both, in duration and intensity. We need scarcely adduce the general symptoms of this disease, having already alluded to them on a prior occasion.

The chief, and pathognomonic phenomenon, is *effusion within the articular cavity*, and rapid change in the contours of the joint. From the physiological character of the structure, effusion, should, *a priori*, be expected, as clinical observation substantiates it.

To speak of a *dry joint* in these affections is an absurdity. The most insignificant irritation of the synovial lining, is attended with *copious secretion* of a fluid, with the peculiarities of synovia. The higher grades may not exhibit the same quantity of morbid secretion, but enough to give definite fluctuation. The liquid is of a more plastic nature, contains blood corpuscles, flakes of fibrin, fat globules and epithelium and becomes early contaminated by the organized elements of pus. To a certain extent the composition of the synovial fluid may still be recognized by the abundance of alkalies and the soapy feel.

In the highest grade of synovitis, the synovial lining, is as you are aware, converted into a pyogenic membrane, and presents the structure of granulations, as stated in the preceding section of our discourse. Under all these conditions, there is more or less morbid effusion.

The dryness of articulations cannot be denied, but it is noticed in conditions of a different character, and independent of inflammatory affections of the synovial lining. Thus, for instance, it complicates progressive deformative arthritis, which originates in the articular faces of the bones and though the synovial membrane may gradually be compromised, it is affected in such a manner as to destroy its character as a secreting structure.

In white swelling, the synovial membrane sometimes presents the peculiarity of dryness, but from anatomical changes of a pulpy character, not the result of direct inflammation.

In pure synovitis we never observe consecutive intumescence, infiltration, or hardening of the surrounding tissues, and never to such an extent as we find it in diseases of the periosteum, and the osseous structure, unless indeed the latter have become involved.

In the more active forms, there is intense pain within the whole joint, with consecutive febrile excitement; but reflex pains are moderate, and the spastic oscillations never very intense. In the lower grades of synovitis (Hydrarthrosis), these symptoms are entirely wanting, and the patient suffers scarcely any other inconvenience, than the effusion within the joint would naturally occasion.

The affections of the periosteum and of the epiphyses, are attended by a widely different group of symptoms. The beginning of these diseases is *very insidious*, and their development so slow as to require months to assume a noticeable form. But little pain attends the initiatory period. The whole trouble marks itself *as weakness* of the limb, dryness and *stiffness* of the joint, with inability to use the extremity in the morning. For a time the contours of the joint suffer no change; and if there be any fulness at all, it is more generally diffused, and extends beyond the limits of the articulation. There is no discoloration of the integuments, though there is frequently that *waxy whiteness*, the result of œdema; whence the term " white swelling." The latter is often the first symptom which attracts attention. Though the patient may have the sensation of heat in the affected parts, it is not *objective* either to the hand or thermometer. The patient may gradually experience some difficulty in using the articulation to the fullest extent, feel induced to spare the extremity in locomotion, and thus favor certain positions as a source of greater comfort; malposition is superadded only at a later period.

The *advance of the disease* is marked by progressive swelling of the periarticular structures: the contours of the joint disappear, not from effusion within the articular cavity, but from infiltration of the surroundings and therefore no fluctuation can be discerned.

Contemporaneous with the enlargement of the articulation, the original feeling of soreness, increases to aching pain, being augmented by pressure and locomotion; the rest becomes disturbed by reflex pains, and the limb forced into a position over which the patient loses all control. Every attempt to alter the same is attended with aggravated suffering.

When the swelling and firmness of the soft parts still more increase, then the pain assumes a torturing character. The limb attenuates and becomes cooler, whilst the swelling shows but a moderate addition of temperature.

In viewing the affected extremity, the contrast between *the waste* of the limb, and the *general enlargement* of the articulation, with its numerous distended veins, is strongly marked, and it is this form of articular disease, which in times past was designated as *fungus articulorum, tumor albus, and white swelling*. It was thought to be of malignant growth, and amputation its only remedy.

Thanks to the progress of pathological anatomy and the material aid of the microscope, this error of our ancestors has been effectually dispelled.

Now-a-days, white swelling has been recognised as an affection of the articular ends of bones, and their respective periosteum; with subsequent periarticular infiltrations of seroplastic material, with its attending organization into fibroplastic cells, fibrous structure, fat, &c. And surgery offers the means of relief as long as the pathological changes are susceptible of reduction.

The knee joint is most frequently visited with this disease, and it is there one can best study its different phases.

On a former occasion I have assigned the reasons why this malady attacks the knee joint more frequently than any other, and likewise why the disease is more frequently observed in childhood than in adult age: and therefore need not recur to that subject.

I shall now confine my remarks to the discussion of some features that characterize the process under consideration.

One of these points is the extraordinary slow advance of the disease. Some authors think that a low grade of nutrition of the structures primarily involved, offers an acceptable explanation. On close reflection we shall find this view inadmissable, and contradictory to analogy. Nutrition in childhood is more exuberant than at any later period. In the former, maintenance is not the only object of the nutritive process; it is enhanced by growth and developement, demanding more ready supply, and meeting with the most elastic condition of the vascular carriers of that supply. In these advantages the infantile skeleton participates to a higher degree than the other systems of the organism.

Hence from a physiological point of view, we have to reject the advanced theory.

In questioning analogy, we notice facts which demonstrate beyond a shadow of doubt, the prolific character of nutrition in the osseous system of children. Fractures consolidate more rapidly with them than with adults; artificial joints are scarcely ever observed during the period of evolution; if periostitis has laid bare the bone of a child, exfoliation rapidly ensues, and sequestra form much more quickly than at a later period. These facts coincide with the experiments of Flourent and Wagner, and dispose effectually of the before mentioned hypothesis.

In all those cases of white swelling, that I have had the opportunity of anatomically investigating, and they have been numerous, I have observed that there is always, in one or the other condyle, an insular disintegration of the cancellated structure, in which sometimes a small sequestrum is imbedded. Under the microscope scarcely any trace of the vanquished structure can be discerned. The chief element is fat. But in the neighbourhood of this pathological focus, hyperaemia, traces of fungoid granulations, and osteoporosis are noticed. This condition explains satisfactorily, the proximate cause of the pathological changes inconsitent with the active process of ostitis. In some rare instances, however, the healthy portion of the bone surrounds the disintegrated isle with a sclerotic capsule, by which the affected portion becomes, as it were, isolated and rendered innocuous, in a similar manner as foreign bodies encapsule. This pathological condition may not cover all cases which pass under the name of *tumor albus*, but certainly this is the most prevalent.

There is a specimen in my collection, of the lower third of a femur of a young girl not exceeding fifteen years of age. She was admitted to the Brooklyn Medical and Surgical Institute, with all the symptoms of white swelling, comprising the articulation and peri-articular structures; the swelling however likewise involved a portion of the femur. The local disturbances were as intense as were the nocturnal pains, and the spasms of the flexor muscles. The knee was of course drawn to a right angle.

From the history of the case, and the clinical character of the disease, *circumscribed osteomyelitis*, with its termination in abscess was diagnosed, and in view of her reduced constitution, and the copious discharge of matter from the neighbourhood of the joint, amputation was deemed expedient.

The condition of the specimen fully confirmed the diagnosis. There is a large pyogenic cavity at the lower end of the femur, which opens at the posterior aspect of the bone, by an irregular aperture not less than an inch and a half in diameter; in the circumference of which, the periosteum is raised up, and its internal surface covered with new bone. The epiphysis is somewhat loosened from its attachment, and in time would have become separated.

The original focus of the diesease had been obviously limited to the cancellated structure, and rather remote from the joint, but its consecutive effects had extended over the joint, and involved its soft surroundings. There may be still *other exceptions* from the anatomical prototype, but their numerical proportions scarcely affect the statistics.

The adherents of the tubercular theory, may rejoice at this pathological admission of mine, of those insular and circumscribed pathological foci, which they may claim as *bona fide* evidence of tubercular deposit.

I hold however, that pathological detritus, limited to an isolated place, cannot in the eyes of competent judges, pass as tubercle.

If the disease is permitted to spread, it eventuates in perforation of the articular cavity; the formation of external abscesses and fistulous tracts, and the more obstacles the discharge has, the more periosteum will be destroyed, and the bone corroded on its surface.

The protracted development of these phases extends over many months, and often additional injuries are required to accomplish so extensive disintegration.

A lull of all symptoms, is often observed in the like cases, to be followed by new exacerbations. A goodly number recover spontaneously, or by appropriate treatment. These recoveries happen not rarely at the period of puberty, at which time the mode of nutrition of the epiphyses becomes perfected.

In analysing the gradual development of this disease, its preceding cause, (traumatic injuries); the comparative moderate effects upon the integrity of the adjacent osseous structure; we find a more passive pathological condition, a direct necrobiosis of the affected structure, more from want of proper maintenance, than from active and progressive disease. When active symptoms subsequently set in, they are the efforts of the *vis medicatrix naturæ* to eliminate the detritus foreign to the integrity of the bone. Frequently the detritus becomes absorbed, or pervaded with calcareous elements, and thus recovery is attained.

This gradual change of the osseous structure and annihilation of its nervous and vascular endowments, though limited in extent, renders it intelligible why so little pain is experienced by the patient, during the first disintegrating period of the disease. The intense pain that is at a later period superinduced, is evidently connected with the peripheral and active process of osteitis arising in the circumference of the focus. The original disease has nothing to do with it.

The appearance of nocturnal pain constitutes a serious complication and indicates the commencement of suppuration.

The *contraction* of the biceps muscle is quite common, and the result of reflected spasm. The leg is thus held in an angular position to the thigh, and most usually *rotated on its longitudinal axis*, with eversion of the toes. This position goes *pari passu*, with an anatomical derangement of the joint itself. The patella rides upon the external condyle of the femur, and is generally adherent; the internal condyle of the tibia projects in front, whilst the external one recedes.

The contraction of the biceps is exclusively accountable for this malposition, for at a certain angle it acts as a rotator, when not *counteracted* by the simultaneous contraction of the internal hamstrings.

I have but lately exhibited to the New York Pathological Society a specimen of this kind, and the action of the biceps is so undeniably demonstrated, that there is no more room for further speculation to account for the symptoms.

For a long time the mobility of the affected joint remains, if not impeded by the contraction, but when synovitis is superinduced to the original affection, the joint may become obliterated by fibrous adhesions between the articular faces, which may still more impede the mobility, but rarely are there osteophytes passing from one bone to the other, depriving the joint of all vestige of motion. True bony anchylosis is of very rare occurrence, and much more the consequence of penetrating wounds of the joint, and high graded synovitis, than of this form of disease.

Whether the disease originates in the synovial membrane, in the crucial ligaments, in the periosteum, or the epiphysis of the joint, the symptoms apertaining to each of them respectively, will be so blended in their advanced course, as to render diagnostic discrimination almost impossible, leaving the previous history as the only guide.

The pathological conditions of joint diseases vary but little, when suppuration, burrowing of pus, has been going on, and the bones have been disintegrated for any length of time; the symptoms attending those conditions are almost uniform in all such cases. The competent and experienced surgeon may yet recognize the patho-genesis of the original disease, but novices rarely realize differences so indistinct and subtle. Thus, in caries of the joint emanating from synovitis, the articular surfaces are more generally denuded of their respective cartilaginous coverings, but the osteo-porosis does not much exceed the surface; the crucial ligaments are but partially destroyed; the semilunar cartilages partly disintegrated, discolored, and mostly detached. On moving the articulation, crepitus is discernible. If, however, the bone has been the starting point of the disease, the caries of the articular surface is generally restricted to the originally affected locality; and the cartilage is there and thereabout disintegrated. The crucial ligaments are mostly destroyed *in toto*, and crepitus is less distinct.

The clinical character of *hip disease* will now demand attention, on account of some peculiarities in its manifestations. Morbus coxarius is about as good a term as could be chosen and certainly more appropriate than " coxalgia " which applies solely to the pain of the affection.

The first stage of this lesion materially conforms with the same stage of the affections of other joints. The only symptom requiring special mention, is limping. It is most noticeable in the morning, less during the day, and least towards evening; most conspicuous after great exertion, and sometimes absent after a day of complete rest. The duration of this period is variable; repeated accidents and the continuous use of the affected extremity may shorten, and constant rest prolong it.

The so characteristic pain at the knee, may already make its appearance at this stage, but if so, there will be likewise indications of retracted muscles, with which this symptom appears conjointly. This pain has often confounded the diagnosis of the less experienced, without any need; for you may press and squeeze the knee joint as you please, without the slightest increase of that pain, whereas the pressure upon, and movement of the hip joint will aggravate it. The progress of the malady may, at this juncture be arrested, and the patient relieved from further trouble.

The second Stage is characterized by elongation, abduction, eversion and slight flexion of the affected limb at the hip, with lowering of the pelvis, flattening of the gluteal region, sinking of the gluteal fold, and an inclination of the internatal fissure, at, and towards the affected side. The mobility of the joint may either be impeded, or entirely suspended. Adduction is generally impossible.

For the purpose of locomotion, the patient brings the lumbar portion of the spine and the other hip joint into play; thereby easily deceiving the inexperienced observer. In the erect posture the spine exhibits a single curve, of which the convexity corresponds with the seat of trouble. The superior spinous process of the ilium, is depressed when compared with that of the other side, and the healthy member is adducted in proportion to the malposition of its afflicted fellow. In walking, the patient places the latter forward and outward, and drags the other limb after it in a rather diagonal direction. All these symptoms more or less complete, can be ascertained by undressing the patient; dropping a plummet line from the occipital protuberance, walking, and by careful examination in the horizontal posture. If the patient sits down in such a manner as to accommodate the affected member, both pelvis and spine assume normal relations, thus proving that the elongation of the limb does not depend on the lateral declivity of the pelvis, as *Gross asserts.

The chief or proximate cause of the entire group of symptoms rests with the immobility of the joint and the fixed adducted position of the extremity. In imitating them we produce the very same effect.

There can be no doubt that the elongation is but apparent, and not real, as the late professor Rust of Berlin, claims. Nor is there any enlargement of the head of the femur, from either tuberculosis or other causes, to which he ascribes the actual elongation. The sole source of the symptom is hydraulic pressure from existing intra-articular effusions. I was led to this view from the analogous position of the femur and the immobility of the joint produced by experimental injection. Acting on this supposition, I have succeeded in substantiating the correctness of

---

*Gross "Practical Observations." Philadelphia 1859.

my opinion, by paracenteses of the articular cavity. The removal of the intra-articular fluid was followed immediately by returning mobility and the correction of the malposition. This point is consequently settled by demonstrable evidence.

With the apparent elongation of the limb, the structural pain gradually increases, and the reflex symptoms rapidly rise to an intense degree. The nocturnal pains, in this period are more violent and torturing than at any later, and for obvious reasons. Whilst the extremity is immovably fixed by hydraulic pressure, the adductor muscles are nightly agitated by reflected spasms, and kept on the stretch. The limb becomes attenuated and exhibits marked disproportion with its fellow, the constitution, rest, appetite, suffer gravely, and reduce the patient in weight and appearance. The effusion may still be of a plastic and organizable character; sero-purulent, or exclusively pus: may be free from, or contaminated with structural detritus, benign or destructive. Its composition will naturally determine the issue of the case. If the effusion be mild, plastic, benign, free from deleterious admixture, its partial absorption and final organization into fibrous structure may take place, and thus terminate the malady. Or its quantity may lead to a disruption of the capsular ligament, and the escape of the intra-articular effusion into the surroundings of the joint, and there become organised and innocuous. Through similar changes the sero-purulent effusion may pass with the same result.

But if the articular contents are of a destructive character, they may, by macerating and corroding the acetabulum pass into the pelvic cavity through the cotyloid notch, or through the capsular ligament, and will invariably give rise to the formation of abscess, corresponding in locality with the place of perforation.

In the moment the perforation is effected a new series of symptoms appears, and with which the third stage of the disease is ushered in.

The third stage is distinguished by diametrically opposite symptoms. The contrast of the two stages can best be realized by placing them in juxtaposition.

Second stage.	Third stage.
Affected limb.	Affected limb.
Apparently elongated.	Apparently shortened.
Abducted.	Adducted.
Flexed at hip and knee.	Flexed at hip.
Toes everted.	Toes inverted.
Foot fully on the ground	Ball of toes only.
Healthy limb adducted	Abducted.
Pelvis lowered.	Tilted up.

Pelvis projects forward.	Backward.
Pelvis angle of inclination acute.	Almost rectangular.
Nates flattened.	Full and convex.
Gluteal fold lowered.	Elevated.
Internatal fissure inclined to affected side.	Inclined towards the opposite side.
Spine curved on the affected side	Curved towards the other side.
Nocturnal pain very intense.	Greatly diminished.

It will be seen that the third stage is characterised by unmistakeable clinical manifestations, and by so peculiar a gait of the patient, as to be recognised at a distance.

The shortening, adduction, and inversion of the limb, conjointly with the rotundity of the gluteal space, strongly convey the impression of posterior superior dislocation of the femur. This similarity of the two may have led Rust to presume their identity, and ascribe to the action of the contracted muscles the cause of *spontaneous dislocation*. The morbid enlargement of the caput femoris, said to exist (at the second stage) lent a plausible argument to this hypothesis. What was more simple and transparent, than that the head of the femur partially expelled from the acetabulum by its disproportionate size, should leave it entirely, and follow the undue traction of the muscles. This hypothesis of the renowned German surgeon prevailed among the profession; spontaneous dislocation was henceforth a settled fact, against which but heterodoxy could raise its voice. Buehring, of Berlin, if I do not mistake, was the first who took issue with Rust's theory, and attempted to reduce the acknowledged similarity of symptoms to causes widely different from those propounded. In this effort, he derived material assistance from the advancement of pathological anatomy. The question once opened has received a rational solution. At this present moment there are few well informed surgeons who recognize spontaneous dislocation. Nelaton has informed us of a good method to decide the relative position of the femur to the acetabulum. In drawing a line from the anterior superior spinous process of the ilium, to the tuberosity of the ischium, it passes on its way, from one point to the other, the apex of the large trochanter, in the normal position of the femur. It crosses the trochanter more or less below the apex in dislocation.

In applying this test in the third stage of morbus coxarius, you will mostly find the normal relations, or so insignificant difference as to preclude all possibility of dislocation. Irrespective to this clinical fact the morbid condition of these points contradict the assertion of Rust *in toto*. It might rather be said that the acetabulum becomes dislocated, since we often find it extending up, and backward in which direction the femur follows, but true dislocations belong to the rarest occurrences. I have searched in this respect the anatomical museums, on this, and the other side of the Atlantic, without having found more than about a dozen specimens, exhibiting the conjoined evidences of hip disease and dislocation. In this statement I am borne out by other enquirers. It follows therefore, that dislocation is but a rare incident in hip disease, indeed much more so, than might be rationally expected, considering the actual state of the joint in many instances. If dislocation is practicable in a healthy articulation, how much more predisposed must the latter be, when the acetabulum is denuded and enlarged, the round ligament totally destroyed, the head of the femur *diminished* in size, the cotyloid cartilage more or less disintegrated, the capsular ligament broken through &c.; which all tend to facilitate the displacement of the femur. It is thus evident, that the slightest appreciable injury should suffice to bring about a dislocation, but its spontaneity cannot be conceived, and must therefore be denied. On the other hand, it must be borne in mind that the joint being more or less tender, is well taken care of by the patient and protected against incidental injuries.

One of these means is the play of all muscles by voluntary effort to keep the joint at rest, and thus dislocations are prevented, which otherwise might seem inevitable. Wherever dislocations take place, there can be no doubt as to their being the result of some injury or other, however trifling. That much I can at least assure, that I never myself have had the opportunity of observing a single case of indisputable dislocation consequent upon morbus coxarius, and I have had my finger in the hip joint too often to be deceived. If you examine a patient so afflicted, with the aid of anæsthetics, extending the affected limb, whilst at the same time exercising counter extension by placing your foot against the pelvis, you will notice a certain amount of mobility of the joint, but the absolute impossibility of abducting it. In searching for the cause, a firm and unyielding contraction of the adductor muscles will be found, over which the anæsthetics seem to have no influence whatsoever. It is thus in the third as in the second stage, the malposition of the limb is produced by a single cause, and the rest of the symptoms follow as physical necessities. Now, for instance, let us presume the femur held in undue position of adduction and flexion, and the patient attempt to walk, he would yield the pelvis as much as possible for the purpose of relieving the tension of the contracted muscles. The first thing he does is to rotate the pelvis in its transverse diameter, thus approximating the anterior superior spinous process of the ilium, to the insertion of the tensor vaginae femoris. This accounts for the enhanced angle of inclination with the horizon. By turning the pelvis on its axis at the lumbar articulations, the patient favors the former object. If the pelvis remained quite horizontal and the extremity of the healthy side rectangular to the former, the affected limb would necessarily cross its fellow, and locomotion would thus be rendered impracticable. Hence the affected side of the pelvis is tilted up in proportion to the adduction of the affected extremity, the healthy member is thrown out, (abducted) and parallelism is thus achieved. If the pelvis is thus out of position, the spine and shoulders have to adapt themselves to the static changes.

In compounding the effects of these changes in the position of pelvis and femur, we can almost to a nicety, ascertain the amount of apparent shortening, without regard to the so called spontaneous dislocation. The longitudinal rotation of the pelvis will raise the extremity as much as an inch, the flexion of the femur upon the pelvis, another inch, and the obliquity of the pelvis from one to three inches. Thus the limb may be shortened in the aggregate, from three to five inches, an amount never to be produced by traumatic dislocation of the femur upon the ilium.

Most cases of morbus coxarius terminate with the third stage; but comparatively a few advance to the fourth and last stage of the disease, which is a combination of the symptoms of the third, with those of caries, abscesses, fistulous openings and tracts, in the neighbourhood of the joint, local pain, arising from such sources, and hectic fever.

Thus it will be seen that hip disease is characterized more than any other, by a certain immutable regularity and chronological succession of symptoms, which, in themselves, furnish the strongest ground for differential diagnosis. Though the first stage may escape the vigilance of the professional attendant, the second will inevitably decide his appreciation of the growing trouble. The third stage is invariably preceded by the second, and the fourth by the former stages. This, at least, has been my observation in a large number of cases, and I entertain no doubt that it is substantially the same with other accurate observers. The exceptions that may be adduced apertain to cases partly not hip disease at all, partly hip disease of a consecutive nature, and consequently blended with other pathological conditions.

Periostitis in the neighbourhood of the hip joint often produces similarities of hip disease of a most striking character. We may find in connection with it all the symptoms enumerated under the third stage of morbus coxarius, but this difference will always be manifest: that the symptoms of the second stage never preceded that condition. If the joint is not secondarily implicated in those cases there will be a freer mobility of the same, and no crepitus; whilst on the other hand, the femur is enlarged and tender.

Sometimes we meet with malposition of the femur in consequence of Potts' disease, and periostitis of the spine, which may give rise to an erroneous diagnosis. The history of morbus coxarius and affections of the spine is so differentially marked that the mistake may be easily corrected. Eventually, the application of chloroform will suffice to overcome the muscular retractions of the latter, and prove the hip joint to be intact.

We owe to Erichsen's careful investigations, our knowledge of the suppurative affection of the sacro-iliac junction, but the symptoms adduced by that author are so widely different from those of hip disease, that they hardly can be confounded. Eventually the careful examination of the corresponding hip joint must necessarily settle all doubts.

IV.
PROGNOSIS ON JOINT DISEASES.

From the preceding remarks of the discourse we may sum up the following prognostic axioms.

From the collective character of joint affections, we must come to the conclusion that they constitute formidable diseases.

In their respective courses, they are slow and protracted, often of years duration.

In their commencement and development they are insidious, and may have proceeded to considerable disintegration of normal tissue before the patient becomes aware of the impending difficulty.

The restitutive powers of some of the articular structures are of an indifferent character, owing to the imperfections of their nutrition.

In as far as the osseous structure is concerned, recovery depends on the gradual destruction of the affected parts, which of course is necessarily tedious.

In most joint diseases the affected structures undergo changes more or less disqualifying them for the performance of their respective physiological offices, thus either impeding or annihilating the usefulness of the articulation.

The suppuration of articular cavities leads to their perforation, to extensive subfascial burrowing of pus, and not only involves the extremity, but the constitution at large.

Reflex pains and spasms accompanying joint diseases are of the most violent and torturing character, upsetting rest and appetite, placing the very existence of the patient in jeopardy.

Caries of the articular faces may cause so copious a drainage as to gradually bring the patient to hectics, pyæmia and multilocular abscess in the vital organs.

Finally, malposition, deformity, false and true anchylosis may terminate these diseases, and disable the patient for the rest of his life.

All this should be borne in mind when taking charge of cases of this description, and our prognosis should be guarded under all circumstances, however slight and insignificant the cases might appear at the first glance; for the objective symptoms are not a reliable barometer of the actual condition with which one may eventually have to grapple.

Notwithstanding all I have said in this respect, the prognosis of joint diseases is infinitely better to day than it was fifty years ago. The present generation has achieved a clearer insight into the physiological and pathological character of joints than our professional ancestors; it has successfully rid itself of errors, heresies, and notions which obscured the unbiassed clinical understanding of this class of diseases; and since then we have steadily improved in therapeutic efficiency and self-reliance. What was formerly a *noli me tangere*, has become a coveted object of diligent investigation and treatment. And the results of our cherished efforts are in every respect gratifying to the professional pride, and afford reasonable satisfaction to the patients concerned.

It will scarcely be necessary to enter into prognostic details, inasmuch as they may be inferred from the previous section of these lectures, or may be yet especially alluded to under the succeeding heading.

V

TREATMENT OF JOINT DISEASES.

The most important proceeding in this direction is a thorough and systematic examination, comprising both the antecedents of the patient and the present clinical aspect of his disease. In reference to the former, the state of health of his immediate and remote ancestors should be ascertained, as it might possibly affect the prognosis of the case. Next to this is the previous history of the patient, whether he has passed through the ordinary infantile diseases without sequelae; whether the previous state of his constitution and health has been strong and vigorous, or otherwise. It might be as well to inquire into the character of his temperament, mode of living, residence, domestic surroundings, &c., in order to form an approximate idea as to the status and vigor of his system.

The next object of inquiry would be the probable causation of the impending disease. In this respect, gentlemen, I should advise to be searching and persevering, for most parents know so little about it, that we are obliged to sharpen their memory. They will assign the most trivial causes, and harp upon the same with great pertinacity, simply because the true occasion is in the past, and has slipped their memory, whereas trivialities are brought forth because they happened at a time, soon after which the disease assumed form and importance. I have been startled by the simplicity with which even modern writers on the subject, have allowed themselves to be stultified with the most innocent and harmless occurrences, as for instance " sitting down on the grass," or " on a cold stone,"or " having run about a good deal," &c. I cannot persuade myself that such trivialities can constitute legitimate and reasonably acceptable causes of joint diseases, even if they are printed over the signature of a respected surgical name.

In closely investigating further, you will learn that there have been *traumatic influences* of some kind or other, more or less *direct* upon the articulation, and if nothing of the kind could be traced, I would not hesitate in assuming the same, if the previous health of the patient had been untainted with manifestations, which can be justly ascribed to chronic nutritive derangements and a vitiated domestic atmosphere. That a traumatic accident has by weeks and even months preceded the actual disease is no argument against its injury, since we know from the preceding remarks, that more or less time will necessarily intervene between the accident and the disease, to bring about those changes in the structures, which can attract attention. Moreover, it is mostly the local pain and the disturbance in the use of the joint, before any notice at all is taken, and either of them are but mere remote results.

We may then proceed with a general inspection of the patient; his general appearance; as to the present state of his health, and the actions of the respective systems. If the patient presents pallor, general attenuation, and prostration, you may rest assured that the disease has far advanced, and shaken his general health by the incidental reactions upon rest, appetite and nutrition.

The patient should then be undressed so as to obtain a full view of the articulation, and the affected member in general; we ought to note its circumference and position, and compare it with the other extremity; ustitute locomotion, and carefully observe how the joint is used and the limb is put to an account. If the patient should limp, we ought to determine whether the limping depends on immobility or tenderness of the affected articulation, or on malposition, or deficiency in the length of the member.

In fine the patient should be placed on a suitable table, so as to be accessible from all sides, and be put under the full influence of an anæsthetic, that volition may be suspended and the rest of the examination be painless. These preparations I regard as essential, to obtain a full knowledge of the character and extent of the disease.

I do not deem it necessary to enter into the full details of the examination with which you are already acquainted. But a few points deserve special attention. In the first place, we have to ascertain the condition of the bones constituting the affected joints, and find out whether the disease has originated remote from the joint, in the periosteum or in the bone itself. In either case, we shall find by comparison, that the circumference of the bone is increased and the adjacent tissues more or less infiltrated, its surface be uneven, pressure upon it be tender, and by bending the bone, we occasionally find that it has lost its elasticity and hardness. We have next to direct our attention upon the condyles, compare their size, elasticity and sensitiveness with the corresponding condyles of the other limb. Frequent practice will enable us to discern changes which are easily overlooked and ignored by the novice. There is a certain degree of elasticity in the condyles, which is lost by the morbid alterations, even the increased tenderness of the bony structure becomes manifest, though the patient be in anæsthesia. On moving the joint carefully, we ascertain the degree of mobility and the changes that may have taken place in the articular surfaces. Polypiform growths of the synovial membrane may thus be discovered, when they are too small for the touch of the finger. Crepitus would be the evidence of destruction of cartilage; its absence proves nothing to the contrary, as we have learned on a former occasion. If the joint allows an undue lateral or rotatory movement, we may infer that the lateral or intermediate ligaments have become destroyed, and if combined with crepitus, it may indicate that the articular faces have been materially flattened and changed in form. If the periarticular tissues of a joint are largely infiltrated, and the joint itself is either dry or contains but little fluid, we have the more reason to suspect bone disease, and centre our attention upon the condition of the osseous structure. A distension of the articular cavity without induration of the periarticular structures, indicates synovitis.

During the anæsthesia, we can but ascertain whether the malposition is produced by interarticular adhesion or muscular contractions, or both, and, moreover, whether the contracted muscles still retain their expansibility, or have more or less lost it. If there are sinuses about the joint we must try to discover their course and termination, though they may be very circuitous. I have found pewter and elastic probes more available for this purpose than silver ones; and large probes better than the finer ones. In this way, gentlemen, we shall arrive at a clear understanding of our case, and establish a reliable diagnosis as a basis of therapeutic action.

THE FIRST STAGE is the disease but virtually. The affected structures are but in a state of congestion and hyperaemia with incident tenderness, there are no substantial changes as yet, and by at once taking prompt measures, we may succeed in obviating future mischief. The earlier this is done the surer we may count on success. Nay more, I should consider myself justified in treating every injury to the joint as a virtual affection of the same. A few weeks' restraint is nothing in comparison with those terrible maladies that may eventuate from apparently insignificant causes. But with all the precautions imaginable, and with the most appropriate and prompt treatment, we are not always able to prevent the consequences, more particularly if they refer to injuries of the periosteum and the bony structure.

The very first therapeutic axiom in the treatment of joint diseases is *rest, absolute* and *unconditional*, and the next, *proper position* of the affected articulation. The efficacy of these two is greater and more reliable than the entire antiphlogistic apparatus, and they generally suffice to meet the exigencies of the first stage.

The affected joint is to be rendered immovable by appropriate bandages, materials, or special appliances; and if the affection concerns the lower extremity it would be additionally advisable that the patient takes to his bed and thus get rid of the superincumbent weight upon the affected joint. The ordinary way of rendering a joint immovable, is by hardening bandages, by leather, gutta-percha, wooden, wire or light metallic splints, that are adapted to the form of the extremity. If the morbid condition of the joint is not far advanced, so that we may not require to inspect the articulation often, and thus disturb the dressing, stiff bandages are certainly preferable, otherwise, splints should be chosen. The stiff bandages are made by impregnating the outer portion of the dressing with flour, starch, or dextrine-paste, plaster of Paris or the liquid glass. Inasmuch as these bandages are more or less impermeable to the perspiration, it is necessary to first surround the extremity with a well applied flannel bandage, under which the unevenness of the surface should be filled with cotton wool. How the rest is done, is indeed very indifferent, as long as it fulfills its object. Until the bandage is perfectly dry, it would be advisable to fasten a splint to the member. In some instances it may be advisable previous to the application of the bandage, to apply an appropriate number of leeches so as to reduce the hyperaemia and stasis, the effects of which are, however, but transitory. The fixture of the joint should immediately follow. Except in recent injuries, the application of cold is rarely demanded, but if resorted to, it should be efficiently applied in the form of ice bags, for which purpose one part of the joint may be relieved from the bandage and exposed to the action of that remedy.

The position of the affected joint should be such in which the patient is most comfortable and at rest. It is chiefly governed, however, by the tendency of certain muscles to contract, and therefore, should at once be placed in an antagonistic position. If you remember that portion of our discourse in which I referred to muscular contraction, you will know to choose the position which is most appropriate. In adopting the same, muscular contractions and malpositions will thus be obviated. Some surgeons advise to give the extremity such an angle as will be most conducive to its usefulness. We have nothing to do with that object at this juncture; our object is to relieve the disease and thus preserve the entire usefulness of the joint; their advice is in place when the joint is about anchylosing. The straight position of the elbow joint gives more relief than the flexed one, irrespective to the fact that the latter favours the contraction of the biceps and brachialis. And a straight limb bears more vertical weight than a bent one, and may be used to greater advantage in locomotion.

The same treatment holds good in perforating wounds of the joints, with the additional rule that the wound be carefully cleaned, its margins properly approximated and united. In this way I have seen many an incised and punctured wound close by first intention, without any inconvenience whatsoever. Different is it with torn and contused wounds, where the first intention is but exceptional, and suppuration the rule. Immobility and proper position of the joint, are likewise the chief indications here, and should be scrupulously observed, but the dressing should circumvent the wound and leave it accessible to local treatment.

In using dextrine, starch and plaster of Paris bandages, that part in the neighbourhood of the wound would be protected by a coating of varnish so as to render it unimpregnable to the discharge.

I rather prefer to secure the immobility of the joint by wire and metallic splints (tin or sheet iron) inasmuch as they will permit the use of permanent bath, which I consider invaluable in the treatment of such wounds. We owe the introduction of this remedy to B. Langenbeck, to whom surgery is indebted for many and valuable improvements. If suppuration of the joint ensues, you will do the most for the recovery of your patient by giving free vent to the discharge, and by keeping the suppurating surface in a very clean condition. By these means, and eventually by free incisions into the articular cavity, I have saved many a patient.

There is hardly any necessity for medication, unless incidental derangements demand therapeutic interference. The local treatment suffices to check and ameliorate the articular disease; time and patience accomplish the rest. Beyond those local remedies I have mentioned, nothing else is required at this juncture. From painting the articulation with tincture of iodine, I have seen no benefit; and fly blisters interfere with the fixture of the joint, cause a needless irritation to the patient, and sometimes give rise to reflexed muscular contraction, as I have seen.

IN THE SECOND STAGE the indications of treatment become more diversified. The pathological character of this period is expressed by structural invasions of a more decided nature; by more copious infiltrations and effusion within the joint; by reflexed pain, muscular spasm and consequent malposition; and, in fine, reactive disturbances of the constitution.

If the patient has been properly attended to at the first stage, the disease will but rarely advance to the second, and if the local affection was of a nature that could not be checked in its advance by due precaution, the second stage will be at least materially mitigated by the previous treatment.

Assuming, however, that the patient comes under your charge with the full pathological and clinical force of the second stage, the same remedies and appliances commend themselves, for *rest* and *position* are the imperative axioms whilst the disease is in active progress. In this stage the antiphlogistic treatment is resorted to in vain, as long as rest and position of the joint are disregarded, and the limb permitted to bend, rotate, or assume any prejudicial posture. Nay more, the antiphlogistic remedies even fail to give the slightest relief or to alleviate one single symptom; my own personal observation has decided this fact conclusively, and I do not entertain the slightest doubt that other surgeons have met with the same negative results. But in securing rest and position to the affected articulation, we almost instantaneously give relief to our patient, and initiate progressive improvements. Having done this it rests with you whether you deem local depletion and the application of ice or narcotic fomentation additionally necessary. I have but rarely and I may say but exceptionally needed them, although I mean not to deny the fact that the distended capillaries may temporarily and usefully be depleted by leeches, wet cups and scarifications; the effect of which you have, however, to render permanent, by means of which I shall soon speak.

If the affected member has already been placed in malposition, you have promptly to reduce the same to insure articular rest. This should be done under the full influence of anæsthetics. I consider chloroform better than ether, and equally safe. If I stated the number of chloroform applications that I have made with complete safety, it might be considered as grandiloquy, and as a slur upon professional brethren who have had the misfortune of meeting with fatal accidents. My mind is free from any such intention: I simply state the facts. Yet I cannot divest myself of the impression that many accident cases might have been obviated by the use of a proper and reliable article, by descrimination of patients, and due care by the administrator.

Of all the chloroform offered for sale in the market, I deem that of Dr. Squibb of Brooklyn the best; it is always of the same purity and specific gravity, of the same physical quality and physiological action, and I use it with perfect confidence.

The mode in which I administer chloroform is very simple, although, perhaps, not economical. I form a coarse towel into a short and wide funnel, with an inch opening at the apex for the free access of air; and look more upon the action of the lungs as indicative, than upon that of the heart. At the very moment that the thoracic respiration ceases, and the diaphragmatic suction prevails, I suspend chloroform inhalation, whether the patient be under its full influence or not. This seems to be the margin of its legitimate use, beyond which the danger commences.

Patients addicted to the copious use of alcoholic liquor, and those that present a leuco-phlegmatic, bloated and hydraemic appearance, are not fit recipients of chloroform; nor would I deem it safe to administer it to patients with a weak and flat pulse, in whom the propelling power of the heart is more or less impeded by the fatty degeneration of that organ.

It has been my fortune almost always to be assisted by reliable and experienced men who watched the effects of the chloroform, and did not divide their attention by looking after the operative proceeding. In a few instances I came near losing my patient by chloroform, and averted the fatal catastrophy only by noticing the impending danger in time. But these mishaps were clearly traceable to that carelessness which arises from the divided attention of the assistant.

The patient being under the full effect of chloroform, we now proceed to reduce the malposition, and bend the limb either in the opposite or intermediate position from that in which we found it. If we meet with resistance we have to overcome the same by a legitimate effort of physical power. I would not hesitate to break up inter-articular adhesions if they offered opposition. If intra-articular effusion opposes the reduction of the malposition, I would certainly perform paracentesis of the joint. If muscular contractions are in the way, I would resort to myotomy or tenotomy.

There are authors who oppose every and all interference with the position of *inflamed joints*, as downright meddlesomeness, and as reprehensible surgical practice, and advise *the reduction of the inflammation* as the preliminary step. I apprehend that their advice is actuated much more by traditional fears, in interfering with inflamed articulations, than by experience.

Unless I were permitted to adopt that plan, I would decline all responsibility attached to the treatment of any joint disease.

I have already stated that antiphlogistic remedies have very little effect upon the inflamed structure of a joint, and none whatever if the articulation is permitted to be disturbed in its needful rest, by the jerks of the patient, or the spastic oscillation of irritated muscles.

If under such circumstances, and under the purely antiphlogistic treatment, the disease becomes arrested, it is in spite of, and not by virtue of such treatment, and probably has been protracted thereby. I could prove this by uncountable cases, and produce the individual patients to prove the facts by their own stories. But such evidence is scarcely needed to gentlemen whose own ore of experience will furnish them with sufficient affirmative facts.

No one will deny the beneficial results of relieving an inflamed articulation of its morbid product, provided that the process of removing the same does not entail additional danger. Mr. Barwell does me the honor of eulogising the operation which has benefitted so many of his patients.

That the operation, if properly performed, is harmless, I shall prove to you on a future occasion.

The division of muscles for therapeutic and orthopœdic purposes in joint diseases has met with an unfair adjudication. Barwell, Davis, Prince and other writers on the subject, are *in toto* against this operation; they hold that extension is quite sufficient to control the spastic affection of muscles agitated by the reflexed effects of joint diseases. My experience in extension in the affections of joints is certainly not inferior to any one of these gentlemen, and perhaps not inferior to them collectively. I say so with due respect to the literary merits of these authors. And I can bring forth, if required, the very proofs of Dr. Davis's error by cases which he had treated by extension for months in succession and in his very establishment, without subjugating the muscular resistance.

Need I state to you that I have availed myself with avidity of all suggestions and means promising aid and comfort to this class of my patients? And it would surely be a source of gratification to me if I could consistently and truthfully acknowledge my professional indebtedness for information, valuable or practically useful. As it is, I am impelled to state, that I have derived little or no benefit from extension *per se* in the treatment of progressive joint diseases. Whatever benefit I have derived from it at all, is unquestionably due to *its collateral effect upon fixing the affected articulation*.

The collective experience on this question I can sum up in the following aphorisms.

1st. Extension cannot part the inflamed articular surfaces, for which it has been erroneously designed by its author.

2nd. Powerful extension is perhaps the promptest remedy against an ephemeral muscular spasm, as every one has experienced with himself if he has happened to be suddenly attacked by spasms of the muscles of the calf, but it cannot be relied on in persistent spastic agitations of the muscles.

3rd. In many instances, extension will not only fail to relieve the spasms, but will re-act unfavorably upon the violence of the existing joint disease, if persisted in.

4th. The division of the contracted muscle is the surest and unfailing remedy.—

The most violent periods in the course of joint diseases I have observed, in consequence of keeping a retracted muscle on the stretch, and nothing short of division would give relief, though many things and the most stringent antiphlogosis were vainly tried before.

It is indeed a most egregious error to assume that the division of contracted muscles is merely of mechanical importance; in some, as yet physiologically unexplained, manner do the contracted muscles relate to the existing joint disease. The retractions never appear before the disease has advanced to a certain degree of violence and structural invasion, and unless overcome in an effective manner, they increase to actual contracture. In all these cases the disease is necessarily protracted, and when at last it subsides, the contracture remains though its original cause has disappeared. On the other hand, the original joint disease may be reproduced after years of extinction, if the contracted muscles are unduly and persistently extended. Some cases of this description are but too lively in my remembrance, and my experience on this subject is too dearly bought to be ever forgotten.

From all this it follows that certain muscular groups stand in vital relation with certain joints, one actuating and irritating the other through the same source of nervous supply. Hence the division of so contracted muscles has a vital bearing on the status of the joint, aside from the me-

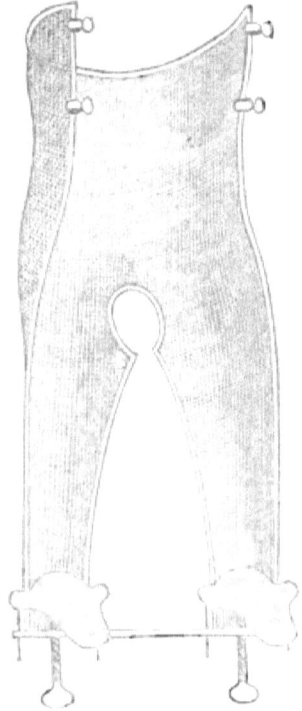

Fig. 1.

chanical relation. In this view we have to judge the therapeutical character of the operation. Dieffenbach already suggested the *antispastic effect* of myotomy and tenotomy; I not only accept his view as correct, but from experience, I am justified in enhancing the same, that in joint diseases at least, it is the most reliable, prompt and unfailing antiphlogistic.

Having suggested and practised myotomy as an antiphlogistic, it is but natural that I should spread before you the grounds on which it stands. The way in which I came to the knowledge and appreciation of this remedy, was simply this; acting on the conviction that rest and position were the two great axioms in the treatment of joint diseases, I had to dispose of muscular resistance as best I could; and often not being able to get rid of it by any other means, I resorted to division. The effects of the division upon the arrest of the joint disease being strikingly beneficial, I gradually included the same as a remedial agent. A practice of fifteen years duration of this operation entitles me to a vote on its merits.

More than in the first stage, rest and position of the affected joint are requisite in the second; and it is in this where special apparatuses are profitably resorted to, to accomplish so important an object. In hip disease, my wire apparatus has not yet been exceeded by any later invention, I place it before you for inspection [fig. 1]. You will see that it consists of a heavy wire frame, which is so covered with wire webbing as to fit the posterior half of the body, from the axillary cavity to

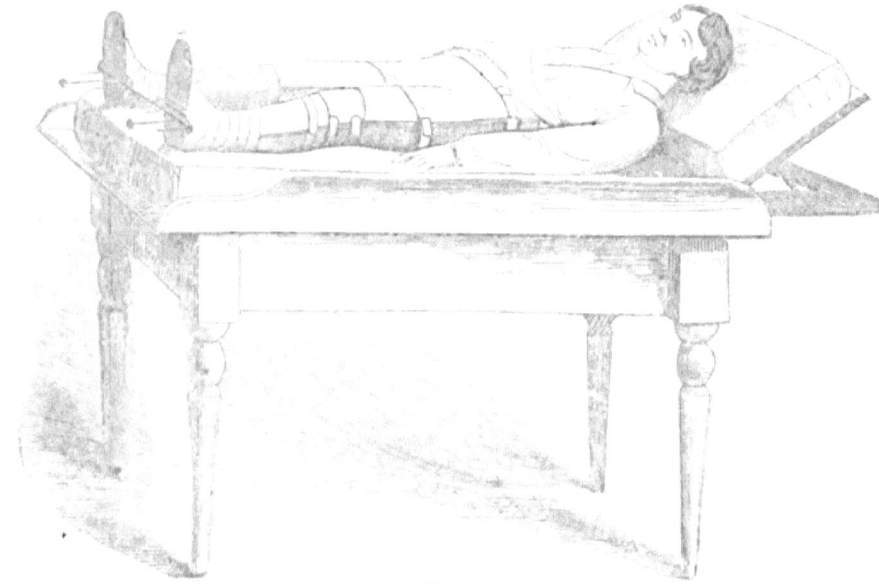

Fig. 2.

the sole of the foot. There is an opening for the anus; the foot boards move by a screw and bolts. To protect it against the corroding influence of urine and fæces, that part of the apparatus most exposed, should be thickly covered with varnish. The average price of the same for children is fifteen dollars currency. In using the apparatus, (fig 2) you have to line it with cotton or other wool or tow, and whilst the patient is under chloroform, you place him in it, and fasten by means of flannel bandages, body and limbs, so securely as to insure his position. If you should desire likewise to apply extension, for greater security of rest and position, you may apply longitudinal and circular strips of stout adhesive plaster, and fasten the former to the foot board.

Some writers, among them Mr. Barwell, have challenged the originality of this invention, and boldly pronounced it a copy of Bonnet's wire apparatus. I apprehend that Mr. Barwell has seen neither, otherwise he could not have come to so inapplicable a conclusion. I have never claimed the introduction of wire into surgery; that point is conceded. Bonnet's apparatus is a clumsy and unwieldly contrivance, produced for no other purpose than to raise the patient by means of pulleys, in such a manner as to obviate painful jarring; my apparatus is an improved Dzondi-Hagedorn where direct extension can be exercised, whilst the counter extension rests with the healthy extremity on the same principle which we employ in having our boot pulled off.

I leave it for you to decide, whether the mode of extension commonly employed in hip disease, offers the same advantages as my apparatus.

In this, position and rest are insured; the patient can pass his fœces with perfect ease, by raising the lower end of the apparatus, and placing a bed pan under it. You can carry the patient from one place to the other, put him in a carriage, draw or drive him into the open air, and thus meet all the objections that have been raised to confinement.

In the other mode, the extension is a fixture of the bed, but what is still worse, it allows the patient to accommodate himself to the position, so as to render extension nugatory. I have seen the patient turn right around, with the perineal band, and accomodate himself so ingeniously that the malposition became as bad as if there had been no restraint whatever.

Davis, Vedder and Barwell, have successively suggested *portative extension apparatus* to obviate the confinement of the patient. The honor of the original suggestion is entirely due to Davis, and the merits of the same ought to be liberally accorded to him, for it certainly has broken the ice of the scrofulous heresy, and paved the way to the rational ideas of therapeutics, which *had been previously advanced*, but disregarded up to that time. Sayre, though strictly speaking, but an exponent of Davis, nevertheless deserves some credit for the adroitness with which he has propagated and popularized the instrument, which seemed to have been an elephant in the hands of the inventor.

Davis's instrument as improved by Sayre is here shown (fig. 3.) But all the before named apparatus are at fault in one essential point: they neither fix the affected joint, nor do they prevent the adduction of the extremity. The amount of extension exercised by them is, moreover, very insignificant, and if it was fifty times as much, it could not separate the articular surfaces of the hip joint, as is erroneously claimed by their respective authors. Besides they depend on adhesive strips for their fastenings, which do not stick well in

Fig. 3.

cold weather, and easily slip in warm. Sayre's modification to circumvent the affected extremity with a semicircular addition at the lower end of the instrument, so as to gain two purchases and two fastenings, was an acceptable improvement in the adjustment, but no more.*

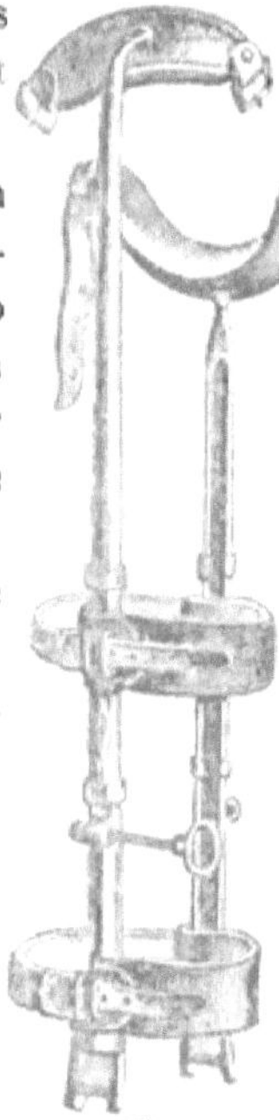

These deficiencies in the mechanical construction of portative apparatus, have obviously induced Andrews of Chicago to fasten a straight steel crutch to the boot, allowing shortening and elongation. In appropriating thus the foot for extension, the tuber ischii for counter extension, and the screw as the moving power, he happily supplied a desideratum and got rid of the annoyance and insufficiency of the adhesive strips.

I had seen nothing of Andrews' very acceptable improvement when I constructed the apparatus which is now before you (figs. 4 and 5). From this to that which I now use, was but one step (figs 6 and 7), it needs no description or explanation, its construction speaks for itself. Not knowing the chronological priority of either Andrews' or my appliance, I will concede with pleasure this honour, if such it be, to my diligent co-labourer on this field of surgical culture.

Fig. 4.

My instrument affords both efficient extension in a vertical line, and complete fixture to the joint, wherein lies its chief usefulness. For two years I have had it in use, and it has given me the fullest satisfaction, in promptly responding to all the indications that can possibly be realized by such a contrivance, and above all it has guarded against the re-shortening of the adductor muscles once divided, which so often happened in my practice, when I used Davis's, Sayers's, and Vedders's apparatus.

That of Barwell, I know but from its illustration; I have never seen nor used it, and forego an opinion on its merits.

With all advantages that may possibly accrue from my instrument, I must warn against its premature use at the second stage, unless the disease has substantially subsided, and 'you intend only to follow up the results of your treatment by its application; the superincumbent weight is too much for an inflamed hip joint, even when supported.

---

* The latest contrivance of this kind is that of Dr. Taylor, of New York. He needed not to have gone to the expense of a patent (!) because it offers no superior inducements and is not likely to be employed by any one else.

To secure the rest and position of the knee joint: I generally prefer metallic splints to stiff bandages. You can handle them better without jarring the joint; you can leave a part, or the entire joint free, for observation and local appliances, and lose nothing in the mechanical effect; you can take them off and re-apply them with the greatest ease: you can combine extension with them, give it inclined plane, &c., and thus secure all the advantages for your patient that could be desired. I generally keep a set of these splints on hand, so as to be prepared for emergencies. The price is but trifling.

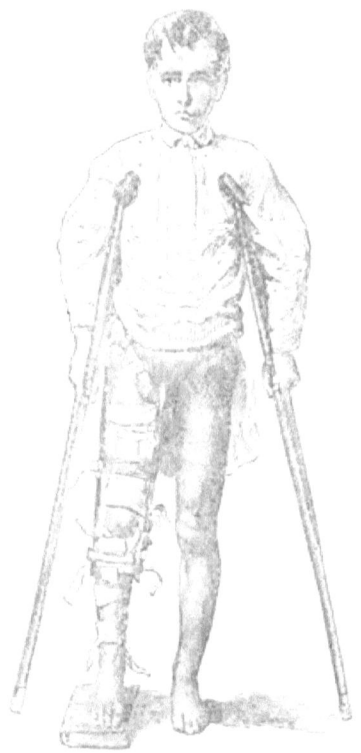

One is a simple gutter splint (fig. 8) for simple cases. The other has a semicircular deficiency at the knee joint to expose one or the other side (fig 9). The third consists of two splints joined by intermediate iron braces designed to leave the knee joint entirely free. (fig. 10)

By drawing bandages from one side to the other across the knee, a moderate degree of anterior pressure may be exercised. If the patient has so far recovered as to resume locomotion with safety, a portative apparatus of an approximate efficacy, should be substitued for the metallic splint. For this purpose, stiff bandages, leather or gutta percha splints, or a special contrivance (figs. 11 & 12) would equally answer. The last consists of two braces along the limb, three or four bands, with a knee cap made of buckskin. If the patient's limb is much attenuated and cylindriform, it would be an improvement to connect the apparatus with the boot, so as to prevent slipping.

Sayre has introduced, for the purpose just mentioned, a portative extension apparatus for both knee and ankle joint, with a view of parting the affected articular surfaces, and thus alleviate pressure upon one another. My belief is that such an object is unattainable by any mechanical contrivance, and moreover superfluous.

In placing an affected joint in such a position as to have the largest possible contact of the articular surfaces, we at any rate diffuse the pressure, if it actually does exist. Sayre's knee apparatus can only be used when the limb is fully extended.

Fig. 6.

In order to perform paracentesis of an articular cavity, the rule ought to be observed, to place the joint in such a position as to drive the liquid to the most accessible spot. At the hip joint this is at the posterior circumference of the acetabulum. The glutei muscles being attenuated, we generally succeed in discovering fluctuation at that particular place. Whilst the surgeon is about inserting the trochar, an assistant takes hold of the affected extremity, and rotates it inwards, which gives the greatest distension to the posterior wall of the capsule. This manœuvre not only facilitates the entrance of the instrument, but likewise the exit of fluid, and prevents the entrance of air.

At the knee joint we have to procure first a straight position, which drives the entire liquid into the anterior portion of the joint. By means of a tightly applied flannel bandage, commencing at the toes, we obviate œdema; the joint is then surrounded with stout adhesive straps, from the tuberosity of the tibia, to beyond the patella; the unevenness of the joint being previously filled

Fig. 7.

with graduated compresses or with cotton. Thus the liquid is driven to the cul de sac, where it is easy of access.—That place in the cul de sac between the duplicature of the vagina femoris and the tendon of the biceps, is most available, there being no muscular structure interposed. Having thus well prepared the articulation, you will easily enter with the instrument, and the liquid will rush

Fig. 8.

out through the canula with great velocity: by moving the finger across the distended portion, you still more facilitate its exit, and with the same finger close the wound, while the other hand withdraws the canula.

I have thus in numerous instances entered the articular cavity, and repeatedly the same articulation, without having caused in a single instance reactive trouble, scarcely ever failed to give instantaneous relief to the joints, although in many cases but temporarily.

Fig. 9.

This is the same proceedure which I invariably adopt in the treatment of hydrarthrosis, and which has proved in my practice a very reliable method.

Puncture of the joint, in these cases, has been unjustly abandoned by the best surgical authorities, (among others, Nèlaton) who considers it dangerous, inasmuch as there is not sufficient centrifugal pressure of the

Fig. 10.

liquid, to prevent the entrance of air, for he states most emphatically that the inter-articular fluid runs out slowly and never entirely. By the

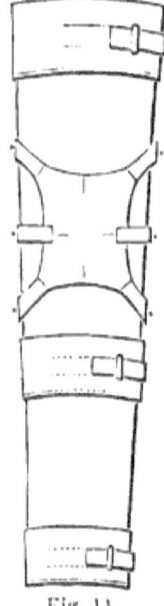

Fig. 11.

plan just advanced we overcome all difficulties and dangers, thus one of the objections may be considered disposed of. The other concerns its efficiency; in this respect, I can but state, that with the exception of one single case, I have radically relieved twenty-seven cases; one by three, two by two, and the balance by one puncture. Of course I have continued compression of the articulation for some weeks after the operation. All the cases operated on were protracted ones of not less than three months, and the majority of more than a year's standing.

This plan, then, compares very favourably in point of dispatch and efficacy, with any other I know of, and certainly is not as hazardous as the injections suggested and practised by Bonnet and Nélaton.

Compression of affected joints is one of the most estimable auxillaries in their treatment, and should be resorted to wherever it is practicable; but when resorted to, it should be thorough and decided. Whether the substance employed for compression has any additional virtue, and whether, therefore, porous or impermeable substances should be used, I am not as yet decided; my experience is almost entirely confined to the use of adhesive plaster spread on Canton flannel, on account of its pliability and durability; and I have been satisfied with the usefulness of these substances.

When, in spite of this treatment, the disease should advance, the articular cavities become more and more distended, and the tendency to disruption is manifest, then the question of free incision arises.

Gentlemen, I am most anxious to put my views on this question so definitely on record, as to leave no doubt as to their bearing and meaning: therefore, I wish to be understood. *First.* That I do not advise nor practice any meddlesomeness with joints at all, unless the strongest indications prevail. *Second.* A moderate quantity of liquid within the articular cavity, whether this liquid be essentially synovia, or plastic or

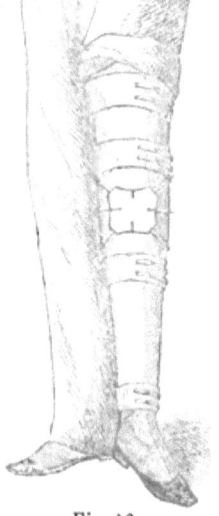

Fig. 12.

purulent effusion, is no indication *per se*, to puncture a joint, for the two former liquids may readily be absorbed and got rid of, and so may pus by previously undergoing a fatty degeneration. I have met with such cases, and but lately the joint of one of my patients opened in the middle of the thigh, from which I could squeeze a large quantity of pus, fragments of cartilage and other detritus, which had for months painlessly occupied the joint, and had completely undergone fatty degeneration. *Thirdly.* I puncture the articular cavity if the effusion is progressive, the distension of the joint very painful; and for the purpose of reducing an existing malposition, provided the latter depends in part or *in toto* on the presence of intra-articular effusion. *Fourthly.* I open affected joints by free incisions, when progressive suppuration of the internal articular surface exists, and threatens disruption of the capsular apparatus.

If I am not mistaken, my esteemed friend, John Gay, Esq., of the Great Northern Free Hospital of London, has first claimed the legitimacy of this operation, and received a goodly share of abuse for it. I have to offer but a few remarks on the usefulness of free incisions. The very essence of surgical wisdom is to imitate nature, and to avail ourselves of similar means for certain purposes. In suppuration the joint is first distended to its utmost capacity by pus, and then spontaneously opened, and the matter forced into the adjacent tissues. The ordinary place of perforations is near the bone, sometimes in part below the periosteum, mostly under the respective fasciæ of the extremities, into the interstices of the muscles, and along the bone; additional destruction is thus caused.

If a joint disease has acquired this character, the joint, as such, ceases to exist: all the structures constituting the internal surface undergo pathological changes, which mostly admit of no reconstruction; the articular cavity is simply an abscess, and should be treated as such. The old surgical axiom "ubi pus ibi evacua," has received its qualification by modern surgery, but its full sway must be recognized, whenever the abscess manifests its tendency to spontaneous opening. For if we have to choose between the alternative of spontaneous perforation, and its undesirable sequelæ, and free incisions,—no surgeon can hesitate in his preference. Sometimes it might be advisable to puncture the joint, and even repeatedly, with a view of obviating the danger of spontaneous disruption; but if the latter presents itself in unmistakeable signs, we should not hesitate in changing the articular cavity into an open abscess, and give free vent to its contents. Hancock, of London, claims exsection of the joint as preferable to free incisions, being more efficacious and less dangerous. There is some conditional truth in this proposition, well deserving con_ sideration. If you freely open a joint and find pathological changes, beyond those of simple suppuration, as for instance, extensive caries; the sequestration of a bone; the partial or total destruction of intra-articular ligaments and cartilages; in fact, changes that would require many months to overcome, exsection of the joint would be infinitely preferable, in such case the free incision would be the initiatory step towards it. On the other hand, if the joint is in a condition of simple suppuration, so that the closing up of the articular cavity by granulation might be safely relied on, the free incision will suffice. In fact, both are distinctly different remedies for distinctly different purposes, and one cannot be substituted for the other.

Having laid down the general principles for the second stage of joint diseases, we may now refer to a few special points. One of them is the treatment of subperiosteal extravasation or effusion; another, the special treatment of those necrobiotic disintegrations of one or the other condyle, to which I have adverted in another part of our discourse. The management of the former is very plain: a subcutaneous division may give all the needful relief, and stop the impending trouble, at any rate prevent its increase. The other is of a more subtle character, requiring a clearly established diagnosis, settled therapeutical principles, and consistent action. How to arrive at the first I have already indicated, and to render the diagnosis still more conclusive the use of an explorative trochar would be advisable. If we have become thus satisfied of the nature of the complaint, trephining by a small instrument, and the subsequent scooping out of the disintegrated tissue, is the most direct and legitimate remedy. I must, however, confess that I have, but in a few cases, resorted to this operative proceedure, though with marked success; my personal experience is therefore limited, but it would seem the most appropriate and direct remedy when a clear diagnosis can be obtained.

In summing up the treatment of the second stage of joint diseases, you will perceive that I rely exclusively on local appliances with a view of obtaining *first, rest* and *position* of the affected articulation. In procuring these I have occasionally to divide resisting muscles and to puncture joints.

Second.—Compression of the inflamed structures.

Third.—Paracentesis and free incisions in joints when suppuration prevails.

Fourth.—In dividing periosteum, and in removing disintegrated bony structure by trephine and scoop.*

---

* Kirkpatrick, *Medical Press and Circular*, Dublin, Aug. 21st, 1867, recommends the use of escharotics, especially potassa c. calce, for the same therapeutic object, and relates most beneficial results.

In the second stage of this class of diseases, we have often to deal with violent constitutional disturbances, which are more readily overcome by proper local treatment than, by any other devised medication, nevertheless the utmost attention should be given to proper diet and hygiene, which is the more necessary as all these cases are more or less protracted, and therefore more or less bear upon the constitutional vigor.

Now, gentlemen, let us contrast the treatment just described with the measures of the old school. Ours is mild when compared with the barbarous derivatory appliances. Moreover, ours is effective; the other is worthless. By our treatment the joint is placed in a condition of spontaneous recovery. The other proposes to subjugate, by direct means, a disease over which it never had nor could exercise any positive influence. Nor is this all; by applying the actual or potential cautery, new troubles are superadded and new taxation is imposed upon an already overtaxed constitution.

But derivation is not only barbarous, useless, and obnoxious, it is even inconsistent with the very pretensions for which it is used. Supposing tubercular depositions are at the bottom of a joint disease, these depositions are either latent and innocuous, or they act like any other foreign substance in creating circumferential inflammation with a view of eventual elimination. In the former proposition, we know nothing whatever of the presence of those depositions; simply because they give no trouble. If we could possibly anticipate the time when such tubercular depositions would be likely to take place, then derivation might be relied upon as a preventive of the impending danger.

But since we have quietly to wait until the so called tubercular depositions are formed, and until they are undergoing the process of softening and compromising the surrounding structures, there is not even a pretence of reason to employ derivation, just as little as if any other foreign substance was lodged within the precinct of the organism. It is claimed that tubercle is not only without organization, but even, not susceptible of it: derivation can therefore exercise no action upon the tubercle itself; that much must be logically admitted. Can it prevent the disintegration of the adjacent structures, and re-establish their former type? of course not; then what is to be expected from derivation at all?

The progress of pathology has been most fruitful in recognising the existing physiological laws which govern alike health and disease. The most reliable observers tell us that inflammations once set up, will run their course to their termination, whether medication be imposed or not. The idea of bringing a recent pneumonia, bronchitis, pleuritis or a catarrh of the air passages to an abortive end has been so thoroughly exploded that no wise practitioner follows any other than the expectant method of treatment, and Hughes Bennett has earned for himself a lasting distinction in proving that fact by clinical statistics. If you concede the fact you have to accept the inferences, that is to say, if you cannot cut off or shorten the course of a recent disease by any means; what can you hope to do in cases of long standing, in structural disintegrations, and more particularly then, when the cause (tuberculosis) is persistently at work.

It will be equally easy to demonstrate the utter uselessness of derivation in the primary affections of the synovial lining. In the mildest form of them (hydrarthrosis) there is a degeneration of the synovial membrane which Johannes Muller describes as lipoma arborescens, which is fully compatible with the increase of the natural secretion, but in which, however, the absorbent powers seem to be entirely lost. Next you have the so called catarrh of the synovial lining in which, according to Volkman, the epithelium is partly thrown off, partly converted into pyogenic source: there you have morbid secretion and loss of absorption. And if you have to deal with a more parenchymatous suppuration of the membrane, you have no longer synovial membrane, but a luxuriantly granulating and secreting surface, with very doubtful absorbing endowments.

The *restitutio ad integrum* is absolutely conditional to the re-establishment of absorption, and this is a question of time. Can you reach or overcome such a difficulty, by blistering or any other derivant applied to the external surface of a joint? Certainly not; like in pleuritic or pericarditic effusions you have either to tap or patiently wait.

I do not want to enter more deeply into the discussion of the therapeutic value of derivation, heretofore unduly praised and over estimated. All I propose is to make a few hints and suggestions, and leave the rest to your mature deliberations.

In the third stage of joint diseases we have still more to deal with both extensive and continued changes in which mostly all the component parts of the articulation are compromised. In whatever tissue the malady might have started, in its progress it has comprised the rest. Thus in synovitis, the articular cartilages have been exposed to constant maceration of pus, and have suffered those elementary metamorphoses to which I invited your attention on a prior occasion. And when at last they drop off in rags and fragments, the osseous surfaces of the epiphyses are in turn subjected to the same obnoxious actions.

With the progress of their disintegration, the periarticular structures become more or less invaded and gradually manifest conditions very similar to those of white swelling. If, on the other hand, the primary affections of the periosteum and epiphysis proceed to the perforation of the articular cavity, it is self-evident that its lining must suffer appropriate alterations. The third stage is consequently a disease of the entire articulation, and its treatment a formidable object of the healing art.

Notwithstanding the undeniable difficulties of these affections, quite a large proportion of the patients recover with or without aid, and sometimes under domestic surroundings of the humblest kind; whereas others run their course to destruction in spite of therapeutic efforts and hygienic advantages. The reason of this difference is not always apparent. Occasionally the abscess determines where the joint gives way to the centrifugal action of the pus. If, for instance, the pus escapes through the floor of the acetabulum, it spreads over the internal surface of the pelvic bones, by detaching the periosteum, and may eventually make its appearance below Poupart's ligament, or through the ischiatic notch, or between the gluteal muscles. Irrespective to the lesion of the hip joint itself, this condition alone would constitute a frightful disease, sure to terminate disastrously. Similar complications may occur with other joints and aggravate their respective diseases.

The indications of treatment diversify with the complications presenting. Generally speaking the same therapeutic rules come into play at this juncture which have been already detailed. *Rest* and *position*, exercise, even in these aggravated cases of joint disease, their beneficial influence, but the appliances should be portative so as to allow the patient the conditional enjoyment of open air perambulations. Of these the patient is greatly in need to sustain his constitutional standard. The appliances should, moreover, be such as would not be easily saturated and soiled by the discharges. James Startin's suggestion to impregnate the bandages and splints of felt, with an equal mixture of melted paraffine and stearine, for the double purpose of stiffening and rendering them watertight, is certainly deserving of attention.

I have not as yet employed this material, but it seems to me preferable to varnish coating heretofore used.

It is self-evident that the fixture of the joint is an essential disideratum to prevent the corroded surfaces of the epiphyses from grinding upon one another, and thereby give rise to pain and renewed irritation.

The fistulous openings should be maintained and their drainage kept free. This is, however, no easy task, because their sinuses are very circuitous, and dilatation by laminaria or compressed sponge, impracticable. The laying open of the tracts by the knife is mostly of but temporary assistance, incurring loss of blood which patients can scarcely bear. The employment of potassa c. calce (Kirkpatrick) to open direct communication between the articular cavity and the surface, deserves surgical consideration.

Abscesses frequently form in the circumference of joints. Those which are attended with great swelling, pain and fever, and indicate the efforts of nature to eliminate structural detritus, should be promptly and fully opened; those which appear more or less remote from the articulation and cause no local or general inconvenience (cold and consecutive abscess: abscessus congestionis) may be ignored as long as they do not raise alarm by their size and pressure upon important parts. Their contents readily undergo fatty degeneration, followed by gradual resorption. But if they require opening it should be done by trochar with the exclusion of air. The knife should only then be employed when air has entered the pyogenic cavity, and decomposed its contents. In this way septicæmia with its fatal consequences can be averted.

With a view of bringing about a more decided detachment and diminution of the structural detritus, various means have been recommended. John Gay insists upon free incisions into the affected joint; others allege they have successfully employed the seton, and Kirkpatrick favours an opening with his escharotic into the joint and uses it freely upon the osteoporotic substance; and finally exscetion. The two former apply only to superficial and accessible joints, and all four are necessarily followed by copious suppuration. They are therefore but available in well preserved constitutions, and in superficial caries of the articular faces.

It is obvious that no debilitated patient can pass unharmed through so consuming an ordeal. As to exscetion I beg to submit:

I. That if a thick slice is removed from the epiphyses, we approximate the cartilaginous disks fastening them to the shaft, which may thus become involved, protract and even prevent the reunion.

II. That if we comprise the cartilaginous disks in the operation, the extremities become so much shortened as to render the result nugatory, and the artificial leg preferable.

III. That the exscetion of single tarsal and carpal bones is but very exceptionally attended with good results on account of the existing intercommunication of the tarsal and carpal joints.

The arrest in the growth of extremities operated upon by exscetion, as observed by Kœnig of Hanau,[*] is probably founded on error and should not prevent us from resorting to so legitimate an operation in its proper place. The growth is impeded by the previous disease, a fact most probably ignored by that author.

---

[*] Archive of Clinical Surgery, Berlin, 1867.

From these remarks it appears that exsection, as well as amputation, has its defined therapeutic value, and one cannot well be substituted for the other without risk and injury to the patient. I have nothing to do with the technicalities of either operation at this juncture.

Permit me, however, to tender my advice in reference to two points in exsection.

I. Before proceeding with the operation, overcome, if possible, the existing malposition by dividing the contracted muscles. I have mostly taken these preparatory steps and thereby secured perfect control of the subsequent position of the extremity. I owe, perhaps, to the observance of this preliminary measure, the happy results that have attended my operations, more particularly at the knee joint. Whereas some of my surgical friends who neglected it, had great trouble to maintain position, and lost their patients. The supposition that the shortening of the limb is sufficient to relax the contracted muscles, proved, in their respective cases, to be erroneous.

II. I remove with great care and accuracy as much of the synovial membrane, serous slides and bursæ (Bilroth) as are extant and exposed to air, for they will suppurate and materially retard union.

At this juncture the debilitated state of the constitution deserves the closest attention. No medication will, however, be of service as long as the local troubles are not mitigated by a proper course of local treatment.

The amelioration of the articular disease is the most direct way of relieving constitutional reaction. Nevertheless, quinine, iron, cod liver oil and sedatives may be needed to control fever, promote hæmatosis, supply an easily digested nutriment, and secure repose and immunity from pain.

In *morbus coxarius* the principles of division of the morbid periods rest on a different foundation, and accordingly the third stage of that disease is determined by the spontaneous disruption of the articulation and a peculiar malposition of the affected member.

It is of course necessary to ascertain the anatomical and clinical character of the existing malady, to determine the plan for therapeutic action.

If the inflammatory character of the disease still prevails, the appropriate means will readily suggest themselves from preceding remarks; and as readily if caries has ensued. The contracted muscles require division to allow the reduction of the existing malposition. Next, the articulation should be kept at rest by means and appliances with which we have already become acquainted; irrespective to the prevailing state of the joint; being equally beneficial in arresting articular inflammation as preventive to the irritative grating of carious surfaces upon one another.

If anchylosis should thus ensue, it will take place in the most desirable and useful position of the extremity.

Locomotion of the patient renders the use of crutches indispensible, the weight of the body will aggravate the local trouble. Only when the caput femoris shows disposition to slide up and backwards, does extension become imperative. My portative apparatus (fig. 6) answers the indications.

When, however, no improvements in the pathological condition of the joint follow this treatment, when caries and suppuration continue, and threaten the patient with hectic, then the exsection of the head of the femur is justifiable and appropriate.

Fortunately the rational and successful treatment of morbus coxarius, lessens the exigency of that operation, and to this fact we may ascribe the present rarity of its preformance.

Notwithstanding the avowed aversion of French surgeons to this operation, it cannot be denied that it has furnished a fair statistic of success, and that it has saved the life of many a patient, which otherwise would have been lost.

Of the seventeen partial exsections of the hip joint which I have performed in the course of my surgical career, nine were attended by recovery and two are still under treatment.

The limbs have been shortened from one to three inches.

With the exception of one case, the sclerotic tissue formed between the acetabulum and the shaft of the femur, permitted a moderate mobility, and is strong enough to bear the superincumbent weight of the body.

That case concerns a young lady upon whom I operated in the year 1856 when she was nine years of age. Owing to monstrous obesity, the intermediate substance has never become firm. I have seen this patient but lately, she has grown to be a handsome and healthy woman; and I have again had an opportunity of examining into her condition. When she stands on her right limb, the mere weight of her left suffices to bring it to its full length. But if she rests upon the latter, the intermediate substance bends outwards and allows the shaft of the femur to come in contact with the acetabulum, by which the limb is three inches shortened. In this positon she can bear the entire weight of the body upon the affected side. My apparatus gives her the desired support for locomotion, and with it her gait is easy and graceful.

I apprehend that some of the exsections which I have performed, have been under rather unfavourable circumstances, and yet withal the conjoint result is anything but discouraging; some of my patients died of other diseases (two of laryngeal diphtheria, and one of cerebral meningitis) evidently connected with the impoverished state of their respective nutrition.

Though I am not a great admirer of exsection of the hip joint, nevertheless I honestly believe that its performance when warranted by the anatomical changes of the joint, bids as fair a chance of success as the exsection of any other joint. It is scarcely necessary to remove carious portions of the acetabulum unless very accessible, for the nutrition of that portion of the pelvis is unimpaired, and inasmuch as it remains accessible to local appliances, it becomes soon repaired.

In those patients who died after the operation, I invariably found the acetabulum restored to its integrity.

VI.
TREATMENT OF THE SEQUELÆ OF JOINT DISEASES.

The most judicious and diligent treatment succeeds but rarely in restoring the affected articulations to a perfectly normal status. There remains generally some tenderness of the articulation, which shows itself after a liberal use, and on changes of the weather. Besides a certain stiffness and dryness may continue a long time after the disease has become completely extinct.

The treatment of this symptom may be fulfilled with aromatic lubrications, cold and warm douche, flannel bandaging, the longer use of "sole baths," which in Germany have acquired great reputation in these troubles. More than all, *passive* and *active exercises* are best calculated to give permanent relief. Even slight malpositions may be gradually overcome in this way. There are quacks in every country who acquire reputation and lucre from the treatment of these articular impediments, and surgeons may learn from them the undeniable benefit of the use of apparently so insignificant remedies as lubricating frictions and passive exercises. I have myself to acknowledge some practical information from this rather turbid source. Having straightened the contracted knee of a lady patient, and repeatedly moved the same under chloroform without succeeding, I at last gave it up. After some months I again met her, with a perfectly flexible and useful joint, and learned that a female quack had restored her extremity to usefulness by persistent and daily lubrications and passive motions. In the beginning, the treatment had been very painful and almost unendurable; but gradually the pain had subsided. I need not to assure you, gentlemen, that this lesson was never forgotten by me; and I am anxious to impart its benefit to you. If you have no time yourself, I would advise you to employ menial hands, but do not give quackery a pretence to superior skill and practical efficiency.

The passive motions are best commenced with the assistance of chloroform, which will enable us to overcome impediments, without any hazard whatever to the patient. Tenderness of the joint may follow, but will subside with a day or two of rest. The passive motions should then be renewed with or without chloroform, as the case demands, and should be carried on until the desired results are achieved. The patient may greatly assist our efforts by appropriate movements.

If, however, the previous treatment has been inefficient and regardless of consequences, the patient will present more aggravated conditions. The very best treatment is no sure protection against an *obliteration of the articular cavity; but malposition of the joint, may and should always be prevented.*

Anchylosis forms, then, another object of after treatment. Surgery discriminates two forms; the false or fibrous, and the true or bony, to which might be added a third, by bony bands or osteophytes. The first consists of partial or total connection of the articular faces by sclerotic tissue, the second in the bony interposition, and the third forms a partial osseous involucrum of the joint. The false anchylosis results from synovitis, both primary and consecutive; the true from penetrating wounds and caries of the articular faces; and the last from suppurative periostitis.

There is always more or less mobility in false anchylosis, but there is no vestige when osseous material forms the connecting link. When muscular contractions existed previous to the agglutination of the articular faces, the mutual anatomical relations of the latter are invariably changed.

The treatment of anchylosis has always been a cherished object of surgery from Hippocrates down to the present time. Success is, however, but of recent date.

Gradual extension for the purpose of overcoming fibrous anchylosis is an old surgical proceeding and has from time to time found advocates in the professional ranks. Mechanical ingenuity has found a fruitful field for display in the construction of all sorts of instruments; the latest method introduced is that by pulley and weight.

The usefulness of gradual extension in the treatment of fibrous anchylosis, is for obvious reasons but *limited* and *conditional*, and the attempt to substitute the same for *brisement forcé* is a failure.

The anatomical conditions resulting from joint diseases are but exceptionally amenable to that method : it is tedious at best, and frequently so painful as not to be borne by many patients. It's claimed superiority is, moreover, anything but conclusive. Nevertheless we meet with cases in which the elastic resistance of intra-articular adhesions and of the capsular ligament can be but overcome by gradual and persistent extension, and in these it seems to be the only remedy. These conditions we recognize only after unsuccessful attempts at *brisement forcé* and the latter has therefore to precede.

Such cases may be rare and constitute but a small fraction in statistics, but they do exist, notwithstanding their denial.

I possess two specimens of this very character, in my collection, both derived by amputation of the thigh. One belongs to a lady who had contracted fibrous anchylosis of the knee from rheumatic synovitis, aggravated by contraction of the hamstring muscles. Before coming under my charge, she had suffered *brisement forcé* without previous division of the contracted flexor muscles. Violent reactive inflammation of the joint followed the forcible extension, and the latter was too painful to be maintained. The integuments sloughed at the internal circumference of the articulation, and her constitution was so violently shaken that her recovery was placed in jeopardy ; and when, after many months of severe suffering, she had regained her strength, she was to all intents and purposes in *a worse condition* than before the operation. Moreover, the leg was in so high graded a state of hyperæsthæsia, that she could not bear the slightest touch, and the thickened epidermis was peeling off in large patches. Although desirous of amputation, I deemed it my duty to try once more *brisement forcé*. Assuming that the omission of myotomy was the cause of the disastrous failure in the first instance, I divided the contracted hamstring muscles previous to the operation. I met no difficulty in breaking down the intra-articular impediments, but I exerted my entire physical strength in vain in attempting to fully extend the leg. I succeeded, perhaps, to an angle of 160° but could not keep the leg in the same. It would jerk back in an instant as soon as I relinquished my efforts.

Applying in the usual manner, longitudinal adhesive straps, and fastening to the same a weight of fifteen pounds, I tried gradual extension over a pulley. No re-action ensued. The limb yielded but very sparingly to extension, and the improvement during the following fortnight was just noticeable. A second effort was then made, terminating as before. I was certain that the muscles had no part in the resistance, having been thoroughly divided. The patient lost all confidence in her eventual relief, and insisted on amputation, which I dared not refuse; for irrespective to the deformity, the hyperæsthæsia alone rendered her condition insufferable. The examination of the specimen revealed the fact that the resistance was exclusively due to the posterior wall of the capsular ligament, which was greatly thickened and pervaded with copious elastic fibres. Even after I had cleared it of tendons, lateral and crucial ligaments, it was impossible to straighten the joint.

The other specimen refers to a little girl eight years of age, who had two years previously acquired an affection of the knee joint through traumatic injury. When I took charge of the case I found her knee joint in an angular position, and its mobility greatly impeded by intra-articular adhesions. There were some fistulous openings at the internal circumference of the articulation, at the bottom of which bare bone could be felt to a limited extent.

In attempting to perform *brisement forcé*, the resistance of the adhesions was very great, and though I proceeded with great care and precaution, I had the misfortune to produce diastasis of the femoral epiphysis. The limb was again placed in its original malposition and kept at rest, and well sustained by plaster of Paris bandages. No trouble at all followed the unsuccessful attempt, and the epiphysis was in due time found firmly united with its shaft. Though I did not feel inclined to hazard another trial of the same kind, but was prevailed upon by the uncle of the patient, who is himself an esteemed physician, and by the family at large. You may well suppose that I was very timorous in the second attempt, and that I used no undue force. In fact the extension of the limb was effected by straight traction and without using the respective bones as levers. On this occasion I succeeded in opening the angle considerably, without being able to straighten the limb completely. But, as in the former case, there was an elastic resistance to contend with, which reduced the angle at once as soon as the tractions were slackened. Moreover the extension of the limb was accomplished at the expense of a shifting of the tibia backward on the femur, and a slight bending of the tibia and femur. There was no separation of the articular faces. Although I had again divided the hamstring muscles, and again allowed the limb to resume its old malposition, nevertheless the ensuing re-action was quite formidable. The patient being of a very delicate and nervous constitution, could not have endured without succumbing to the violence of the symptoms, and therefore amputation was resorted to to avert the fatal catastrophe. Happily, recovery ensued without any untoward occurrence.

In this specimen the resistance was due to the strength and elasticity of the intra-articular fibrous adhesions, and I was unable to overcome it by any means short of entire demolition of the specimen. In attempting to straighten the same, the epiphyses of both constituent bones were proprotionately compressed and the shafts bent, whilst the anatomical relations of the joint remained unchanged.

It is very evident that from these and similar causes, the extension per force, is not always practicable, and there remains, consequently, a limited orthopædic field for the employment of gradual extension.

When in London, I saw a young woman at the Royal Orthopædic Hospital, who had been successfully relieved by gradual extension, from a fearful distortion, caused by a very thick, and apparently unyielding cicatrix, the result of an extensive burn. Her chin had been literally drawn down and fixed to the chest. She was then still under treatment, but her head stood already erect, and most of its motions were free; the cicatrix was soft and pliable. This startling result had been achieved by persistent gradual extension throughout three successive years.

The anatomical composition of scar tissue is the same which characterizes the fibrous impediments of my cases, and if the former can yield to persistent extension, the latter likewise will.

In preferring this method in any given case, I should advise to remove all and every muscular resistance by previous division. There are some authors, among whom Barwell occupies a prominent position, who oppose, for several reasons, this operation as unnecessary and objectionable. According to their reasoning the contracted muscles are in a state of clonic spasm, which will yield to persistent extension.

I have already exposed the fallacy of this opinion in another place, and proven by theory and practice the inefficiency of gradual extension, in as far as muscular contraction is concerned. But if it is impossible to extend them in more recent cases of joint disease, it is surely impracticable in protracted cases, and after the muscular tissue has been displaced by structural elements devoid of expansibility.

From my experience, gradual extension is absolutely dangerous, being apt to produce fearful and insufferable pain, and reproduce the original disease of the joint.

I am indeed astonished at the self-assurance with which Mr. Barwell claims invariable success. The field of his clinical observation must indeed have been very limited when he never met with cases in which gradual extension gave rise to serious troubles. All his arguments against the division of contracted muscles are, moreover, of a very insignificant nature. Mr. Barwell says the divided tendons of muscles do not readily unite. I deny this assertion as entirely unfounded; if the division is carefully performed and the theca of the tendon respected, it will unite readily and form firm and reliable connection. My experience has been rather the other way, and therefore I have been occasionally compelled to re-divide the same structures.

Next, it is asserted that the divided muscle is so much shortened by the operation as to lose entirely its physiological office. However, how can the muscle lose a function which it does not possess? The division of muscles which had not entirely lost their physiological expansibility, does not permanently destroy it; I have had plenty of proofs to that effect in my practice.

The fact is that most of these muscles are worthless before and after their division, because most patients content themselves with a straight and useful extremity, though the mobility of the interested joint may have been partially or totally lost.

The inefficiency of gradual extension has led to the adoption of a more efficacious and practicable method for the treatment of fibrous anchylosis, known as forcible extension or *brisement forcé*.

Some twenty years ago, Amussat called the attention of the Royal Academy of Medicine to the method of M. Louvrier, and caused a committee to be appointed to investigate its startling results. The report thus elicited from competent surgical judges, presented, that up to that time Louvrier had treated twenty-three cases of contractions of the knee joint; that he employed a rather clumsy and complicated apparatus by means of which he forcibly broke down all resistance and straightened the respective limbs; that the results were but imperfect; that no good form was obtained; that a few had been straightened perfectly and remained so, that in some, posterior subluxation of the tibia had been produced and that three patients had died from operative shock, purulent infiltration and pyæmia. Louvrier himself admitted, with laudable candour, the short-comings of his method.

In spite of the enthusiasm on the part of the younger members of the profession for the new method, it met with but a cold reception among the contemporaneous surgeons of note. But a low therapeutical estimate was put upon it, and at best it was pronounced a cruel measure worse than the trouble it was designed to relieve. Fergusson and Stromeyer were its most determined opponents and disposed of it in not very flattering terms.

If I do not mistake, Dieffenbach was the only surgeon of distinction who not only vindicated *brisement forcé* but had the courage to adopt it against all clamour. He, however, modified the proceeding by substituting the hand for the surgical rack of Louvrier, and included tenotomy and myotomy as preparatory measures.

In a comparatively short time this distinguished surgeon had operated upon 200 patients, and reports the general result in his work on operative surgery, to the effect that he lost but two patients from suppuration and pyæmia; amputation was required in one; in some the limb was improved to a moderate degree, in others anchylosis became re-established. A large proportion of the patients were materially benefited.

Some advancement has this method of treatment received at the hands of Professor Bernhard Langenbeck, of Berlin, but it should be remembered that he had a most powerful aid in chloroform. In his inaugural dissertation, on entering upon his professorship,* he pronounces gradual extension ineffective; the division of the contracted muscles, as performed by Dieffenbach, as superfluous, and even dangerous, by inviting the entrance of air and thus giving rise to suppuration. Louvrier's method is, according to him, too uncertain, and its results removed from the control of the surgeon. The technicism of Langenbeck conforms, in most points, with those of Dieffenbach. The results which Langenbeck attained up to 1853, are compiled in the inaugural dissertation of Philipp Frank.†

In carefully analyzing the results of Louvrier, Dieffenbach, and Langenbeck, and in comparing them with each other, it cannot be denied that Dieffenbach's were superior to Louvrier, and Langenbeck's better than his predecessors. But all of them are certainly imperfect, and by no means satisfactory. Louvrier caused, in three cases, considerable injuries to the knee-joint, and consequently lost them. Of what nature these injuries were I have not learned, nor the reason why they happened in three cases, and not in the remainder. Very likely that they were cases of true anchylosis, and that he fractured the bones, or caused diastasis of the epiphysis, or tore vessels or nerves. The subluxation of the tibia, in almost all the cases of Louvrier, must have been a great detriment to the final result of his treatment. For, in the first place, the posterior projection of the tibia must have, by necessity, compressed the popliteal nerves and vessels, thus materially interfering with the circulation and innervation of the leg. Again, the gastrocnemius was evidently put on the stretch, and the heel prevented from reaching the ground.

---

*Commentatio de contractura et anchylosi genu nova methodus violentiæ extensionis ope curandis. Berolini, 1850.

† De contractura et anchylosi articulationis genu et coxæ; Berolini, 1852.

Moreover, the contracted flexor muscles were so much irritated as to cause serious subsequent troubles. Dieffenbach's method was, therefore, a material improvement. In using *manual* force alone, he protected himself against the error of meddling with cases of true anchylosis, not amenable to brisement forcé, and by dividing the contracted muscles he relieved the patient from the serious consequences appertaining to undue extension. Lastly, in breaking the anchylosis up, by alternate flexion and extension, he obviated subluxations of the tibia. The real merits of Louvrier or Dieffenbach for the advancement of this province of orthopædic surgery are, in my humble judgment, obviously greater than those of Langenbeck. The method of the latter is essentially that of Dieffenbach deprived of the benefit of tenotomy, but favoured by chloroform.

I have the most unreserved appreciation of the great talents and diligence of Langenbeck, but I appreciate truth and clinical facts still higher. About 600 cases of affection, contraction, and anchylosis of the knee-joint have given me ample opportunity for most thorough clinical observations, and entitles me to a participation in the settlement of the important question which is still being discussed by the highest scientific tribunals of Europe, before which Langenbeck maintains his former position.

On the feasibility of *brisement forcé* we all agree. Its superiority over progressive extension can no more be questioned, and its former opponents have been effectually silenced by the overwhelming results of that practice. It has also been clearly demonstrated that the hand is a better mechanical adjuster than the lever and the screw. But for the introduction of anæsthetics, more especially of chloroform, the operation would have been of little practical value. The pain attending it is severe enough to terrify the boldest patient and surgeon. The subsequent sufferings it entails, and the uncertainty of its success, would have driven it again into oblivion. Chloroform and tenotomy assure the future of *brisement forcé*. The former renders it perfectly painless, the latter protects against consecutive effects, which are worse than anchylosis and the contraction of the knee-joint together. I do not dispute that in some instances, simple extension will suffice to overcome, permanently, a moderate reflex contraction. Further, I have observed that a weight of a few pounds fastened to the extremity for a few days will have the same effect. But a high degree of muscular contraction can be subdued by division alone. The name of Langenbeck was sufficient inducement for to follow his directions.

I have tried his method in quite a number of cases, and succeeded, in most of them, in extending the extremity, but as soon as the anæsthesia subsided, the muscles commenced contracting again, or, if prevented therefrom by mechanical restraint, an intense suffering ensued. There are but few maladies that cause so intense agony, and prostrate the constitution in so short a time, as the persistent extension of contracted muscles. I remember, among several cases, particularly one of a little boy, who was brought on from Montgomery, Alabama, with a contraction of the knee-joint. The original disease, synovitis, had subsided two years before. The joint was quite well, and there was no pain felt either on motion or pressure. Moreover, the mobility of the joint was not materially disturbed beyond the impediment of the contracted flexors. Under chloroform only the biceps muscle felt tense, and I divided it. I then easily succeeded in extending the leg, and in securing its position in a straight splint. The anæsthesia had scarcely passed off, when the patient began crying loudly, and very soon the articulation became tender and distended. Inflammatory fever set in, with a pulse of 150. The strongest opiates, the most active and persistent general and local antiphlogistics made no impression whatsoever. The paroxysmal pains suggested to my mind their specific character. On relieving the limb from its restraint, it immediately bent. This was another indication in the same direction, and yet the tension of the remaining undivided flexor muscles was so trifling as scarcely to deserve notice. On the sixth day after the operation, the joint was greatly distended and fluctuating, without the slightest sign of amendment. At that juncture I again placed the patient under chloroform, when, again, all muscular tension vanished, and I had to wait for the subsidence of anæsthesia in order to mark the tendons to be divided. What sedatives and the whole antiphlogistic apparatus failed to effect, *tenotomy* did. Rest immediately ensued therefrom. From that moment improvement commenced, and eventuated in perfect recovery. I could adduce several instances of the same striking and conclusive nature. But one will suffice to illustrate the importance of tenotomy in the treatment of the deformity under consideration. I shall now proceed to delineate the plan which I have adopted, and which I have reason to believe is the mildest, the safest, and certainly the most effective. First, be certain in the diagnosis. Fibrous anchylosis may be easily recognized, for there always remains a moderate degree of mobility at the joint; even osteophytes are not incompatible with mobility, more especially when they arise from one bone, and do not firmly connect with the other. But if both bones are united by osteophytes, there is nothing left of mobility, and in as far as the latter is concerned, there is no symptomatic difference between a true anchylosis and that caused by osteophytes. The previous history of the case alone can give you a clue as to the nature of the anchylosis. From the preceding remarks you may be led to expect osteophytes from previous periostitis, and true bony union from a more structural affection of the joint itself. Supposing, then, that we had either a fibrous or an osteophytic anchylosis, with marked contractions of the flexor muscles, I would suggest, first of all, to divide all the contracted muscles. It will be better to do this six or eight days previous to the performance of the *brisement forcé*. By that time the wounds have firmly closed. No air can enter and give rise to suppuration, and you obviate at least one of the objections raised by the opponents of tenotomy. It is, of course, indifferent whether you use chloroform on that occasion, since but little pain accrues from the operation. Nor do I deem it necessary to give you special advice as to the flexor muscles of the leg, since by extension you can raise them from the adjacent parts, and divide them successively as they present themselves. The division of the tendon of the biceps deserves special mention. The external popliteal or peroneal nerve is in such close approximation to the internal margin of the tendon as to be easily cut through. If this be the case, paralysis of the abductor muscles of the foot and talipes varus would inevitably follow. In order to avoid this nerve, you have to divide the tendon either from outside by dead pressure with a tenotome not too sharp, or by inserting it close to the inner margin of the tendon, and give the edge an outward direction. With all precaution imaginable, I have nevertheless met with this accident in four cases. Yet I am happy to say that the paralysis arising from the inadvertant division of the nervus peronæus, did not exceed six months, the nerve having probably re-united, and thus re-established its full innervation.

About eighteen months ago, I took charge of a young man, who had sustained a serious accident; his right knee-joint having been opened at its outer aspect by a large lacerated wound. The tendon of the biceps as well as the peronæus nerve were demolished for about an inch. The patient has never recovered the action of that nerve.

But even if there be no trace of mobility in the joint, as in complete osteoyhytes, tenotomy should precede *brisement forcé* for reasons requiring no further explanation.

In order to perform *brisement forcé* the patient should be fully under the influence of chloroform. He should be placed on his face, but at the same time due attention paid to respiration, for at that degree of anæsthesia, respiration is very feeble and in the main diaphragmatic. The slightest impediment may entirely arrest it. As soon as the patient is thus prepared, you have the thigh properly fixed by an assistant, and then taking hold of the leg, bend it with a sudden jerk, and then extend it; and so continue to alternate between flexion and extension, until the articulation is quite free.

If there be any rotation of the tibia, it will be advisable to amend that position by re-twisting it in the opposite direction. This done, bandage the extremity carefully with a roller, surround the knee-joint with strips of stout adhesive plaster, and fasten either the extremity in a straight iron splint, such as I have before shown, or adjust the extension with the pulley and weight, as before described. In order to correct the lateral position of the limb, Professor Robert places side cushions inside of the splints before fastening the extremity.

By this plan I have obtained most satisfactory results, and have never had any trouble in producing a speedy and steady recovery of numerous patients. It was never followed by inflammation or neuralgia which other surgeons have complained of; nor did the contraction return, provided all the contracted muscles had been successfully divided. If any of those symptoms should set in, rest assured that the tenotomy is not complete. The earlier you perfect it the better it is for your patient. It is needless to contend against them by antiphlogistics and sedatives; you will effect nothing. Tenotomy is the only remedy.

Brisement forcé is both in appearance and reality a powerful remedy. It overcomes, by main force, all resistance; it ruptures the fibrous adhesions of the joint and unyielding tissues, and can certainly do great mischief if indiscreetly performed. But in using the necessary precautions with physical power, nothing is to be apprehended therefrom. In the large number of my cases I have had but four accidents: one of them was inevitable, and certainly could not be foreseen. The case refers to a youth of about sixteen years. He was tall, slender, and evidently of feeble constitution. Having been employed in a manufactory in which he had to tread a wheel, he had thus acquired an inflammation of his knee-joint, which terminated in a deformity. His leg was bent at an angle of 105°, (Fig. 13), but permitted mobility within an angle of 30° beyond which there was resistance on the part of the contracted biceps and other articular impediments. The patella was moderately moveable. After having divided the tendon of the biceps, I increased the flexion of the limb by a comparatively gentle effort, when, to my surprise, the resistance suddenly yielded.

A few days afterward a slough appeared in the popliteal space, and the suppuration became so profuse as to render amputation imperative. It was then found that the epiphysis of the femur had yielded, whereas the articular adhesions had remained unbroken. (Fig. 14.) The disproportionate strength of the articular adhesion, over the union between the lower extremity of the femur to its shaft, was the proximate cause of the

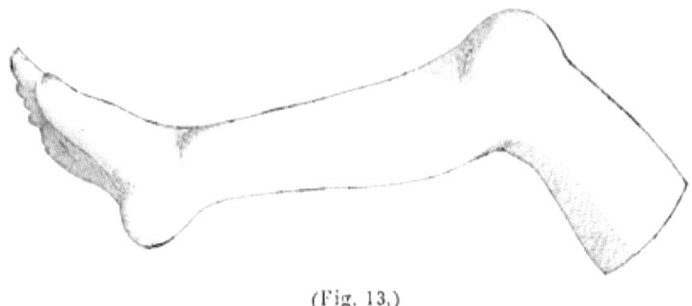

(Fig. 13,)

accident, and certainly could not have been anticipated. A large proportion of my patients have been children in whom the same condition of the femur existed, but with the exception of a few cases, I have met with no accident whatever. In reference to the case just related, I candidly confess that I had not the remotest idea that such an accident would happen at the age of the patient, nor did I or any of my able assistants realize its occurrence. It was in fact the first mishap of this kind, though it has not been the last. The next case happened with a lad from Indiana, aged 17 years. His appearance was equally delicate, but more from rapid growth than any other cause, for the affection of his knee-joint had

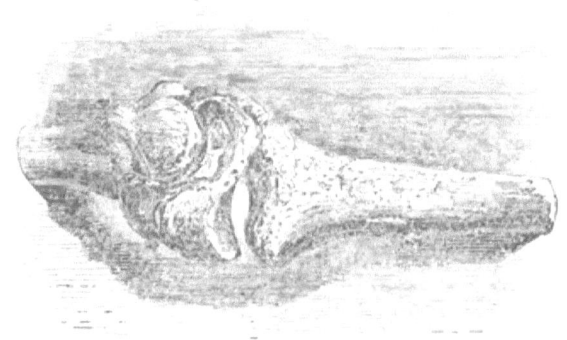

(Fig. 14.)

subsided some years previously. I performed the operation at the office of my esteemed friend Dr. Gaston at Indianapolis, and in the presence of the prominent practitioners of that city. They all can bear witness that I proceeded with great care and precaution, and employed no undue violence.

Nevertheless a diastasis of the lower epiphysis of the femur took place, but no serious consequences followed, beyond the delay of treatment, which has since been resumed.

The other two accidents of this description happened with children; one of the cases I have already adverted to.

These accidents are indeed of no great consequence, provided they are promptly recognized and attended to. The limb must be brought back into the former position, and this position must be secured by bandages impregnated with plaster of paris; in six or eight weeks the union is perfect, and the treatment may then be renewed without further hazard.

It seems to me that these accidents are likely to happen in cases where the intra-articular adhesions are rather tough, and the connection between the epiphysis and shaft of the femur somewhat infirm. The latter may be expected in debilitated and overgrown individuals, and, in such, more than usual precaution is needed to obviate mishaps of this description.

If we consider the small proportion of accidents connected with brisement forcé, and the large number of operations I have successfully performed, they scarcely command our notice. But even this insignificant number of accidents may be reduced by still greater precaution, and during the last two years I have successfully avoided them entirely, and hope to do so for the future. Whenever I have reason to suspect infirm epiphyseal connections, I do not attempt to break up at once the intra-articular impediments, but do so in three or four different times and secure each time the gain by appropriate mechanical appliances. The safest way, however, to break up adhesions of this description is by extension and not by flexion, as I have before advised. The latter is more efficacious but more dangerous in producing diastasis.

In protracted cases of false anchylosis, we are likewise necessitated to repeat the forcible extension several times before succeeding in giving the extremity the full benefit of a straight position, and we may succeed at a third or fourth repetition when the first attempt proved very inauspicious. This is especially the case when peri-articular scar-tissue complicates the mechanical difficulty.

After the brisement forcé has been performed, the extremity should be firmly surrounded by a well applied flannel bandage, with ascending tours, from the periphery towards the interested joint, and the latter with tightly applied strips of adhesive plaster spread on Canton flannel, over which the flannel bandage is continued to the body.

The extremity is then placed in a well adapted and well padded iron splint, and thus secured, kept at rest for several weeks, until the last vestige of soreness of the joint has disappeared.

When the patient is perfectly free from pain or other symptoms, he may be permitted to leave his bed, and walk, but even then the limb should be supported by the same instrument which I have recommended for the after treatment of inflammation of the knee-joint. (Vide Figs. 11—12.)

Most patients content themselves with a straight, useful and stiff knee-joint. But very few insist upon the re-establishment of motion. In this case all those measures have to be adopted which I have detailed under the treatment of stiff-joints. To realize a full share of mobility under these circumstances is a therapeutic object of considerable difficulty, and should not be entertained without due deliberation. The number of cases in which I have succeeded in re-establishing motion is very small, and in two only perfect. If we consider that in most of these cases the articular cartilages and the synovial lining are destroyed, and that the intra-articular fibrous tissue passes from bone to bone, and from wall to wall, we should not be surprised when success attends but rarely these efforts. Moreover, the intra-articular fibrous tissue again rapidly unites with the same articular surface from which it has been torn, and this is an additional difficulty in the re-establishment of free motion.

When osteophytes unite the bones between which the joint is formed, there is of course no mobility, and the firmness of the joint simulates that of true bony union, although the previous history of the case may suggest the character of the abnormal connection. The *brisement forcé* is after all the only safe diagnostic test. Fortunately the osteophytes are not true bony structure, and possess neither its elasticity nor its firmness. These bony splints are rather fragile, and break readily with a crackling sound as if true bone was giving way.

The presence of osteophytes does not in any way interfere with the *brisement forcé* and its ulterior results, the after treatment, nor is it materially affected by them.

In extensive and complete osseous union of the knee-joint, *brisement forcé* is of course ineffective. Rhea Barton's operation alone is calculated to meet the emergency. Although originally proposed for the relief of anchylosis of the hip-joint, its author conceived the practicability of the operation in the same morbid condition of the knee-joint. In 1835, he, for the first time, performed the exsection of a wedge-formed piece of bone from the knee, and the result attained was highly satisfactory. The wound closed in two months, and in five and a half months the patient resumed his avocation as a practising physician.

The second operation of this kind was resorted to by Prof. Gibson, of Philadelphia, and likewise terminated favourably, the patient being capable of walking, without crutches, five months after.

The third operation Dr. Gordon Buck successfully performed at the New York City Hospital, in 1844. The patient subsequently sustained a fall from a ladder and fractured the new union; recovery ensued without any untoward accident.

Since then the same operation has been repeated by Mutter, Bruns, (Tubingen,) Heuser, B. Langenbeck, Reid, Robert, Post, (New York,) and others. As far as I have ascertained, but two cases proved fatal (Bruns and Post;) the balance recovered with useful extremities. The technicalities of Barton's proceedure may be found in every work on operative surgery.

The late Prof. Brainard, of Rush College, has, some years ago, suggested weakening the inter-articular substance by drilling it in various directions through a small wound, and then to fracture the rest. How many operations have been made according to this plan, I do not know, but its application signally failed in a case of one of our most accomplished surgeons, (Prof. Gross,) and a chisel had to be resorted to, which was driven through the bony connection.

A similar proceeding had been proposed by Prof. Shuh, of Vienna, as early as 1853, but did not meet with the approval of German surgeons.

Whether the recently introduced so called osteoplastic operation of B. Langenbeck has been attempted in true anchylosis of the knee-joint, I am equally ignorant, but apprehend that a simple separation of the articular faces by drill or saw will scarcely suffice to give a good form to the extremity, the new bony substance being an impediment; and, therefore, I would prefer, of all the methods suggested, that of Rhea Barton, which has proven itself both effective and comparatively harmless.

The indications for and the technical execution of *brisement forcé* are in most other joints the same as at the knee-joint. But in reference to the hip-joint the operation is subject to some modification, with which I shall now occupy your attention.

Before entering upon the practical consideration of the subject, a short recapitulation of the anatomical condition of the joint, left by hip disease, will not be out of place. Like the knee-joint, this articulation presents the three forms of anchylosis. Of these the true or bony anchylosis is certainly of very rare occurrence judging from the few specimens of this character which can be found in the most complete collections of morbid anatomy. I do not think that I have seen more than two cases during a practice of nearly thirty years duration. Osteophytes are often met with in the neighbourhood of the hip joint recovered from morbus coxarius. Fibrous anchylosis is unquestionably the most common result of that disease, and we find it generally complicated with malposition of the thigh, arising from muscular contractions.

I have had repeated opportunities of thoroughly examining the anatomical status of joints thus changed. In the first place I have found the acetabulum enlarged in a posterior and superior direction, giving it almost the shape of a figure eight; the new accession being the smaller part. The cartilaginous covering of the acetabulum proper had almost entirely vanished, and upon the accessory portion none whatever could be detected. In some instances the femur was riding on the remnant of the acetabular margin separating the two articular segments, and for this purpose had a corresponding groove which gave it an accurate fit.

Of the femur, the head had been entirely lost in every single instance, and the neck more or less shortened.

The intra-articular fibrous adhesions fastened the end of the femur to the articular surface of the pelvis, permitting a slight degree of mobility. The capsular ligament was more or less comprised and identified with the intra-articular fibrous structure, and could only in one case, and to a slight extent, be separated therefrom.

In two instances fibrous bands obviously of a neoplastic character strengthened the connection of the femur with the pelvis. The osteophytes arose from the neighbourhood of the acetabulum, were short and thick, forming no organic connection with the femur and would have offered no impediment to the brisement forcé.

From this short sketch we may arrive at an approximate estimate of the prevailing anatomico-pathological conditions which *brisement forcé* has to contend with.

Buchring was the first who extended the usefulness of *brisement forcé* to the hip joint, and made strenuous efforts to correct the co existing deformities. The means employed by him were, however, so defective that but imperfect results were attained. He already adverts to several cases of failure and disaster; in one he reproduced the original disease to which the little patient eventually fell a victim. And I have to place an instance on record, in which by a fall, *brisement forcé* was effected and subsequently followed by the return of the disease, terminating fatally. The case happened with a lad of Swedish extraction, about sixteen years of age. The original disease had taken its course through several years, terminating in fibrous anchylosis of the joint and malposition of the femur, when the patient was about ten years old. Aside from the existing impediment to locomotion, he had not been troubled for six years, when he fell down stairs and thus forcibly broke the existing adhesions. Violent suppuration followed the accident, and destroyed life by pyaemia. Having secured the specimen (Fig. 15), I had the rare opportunity of satisfying

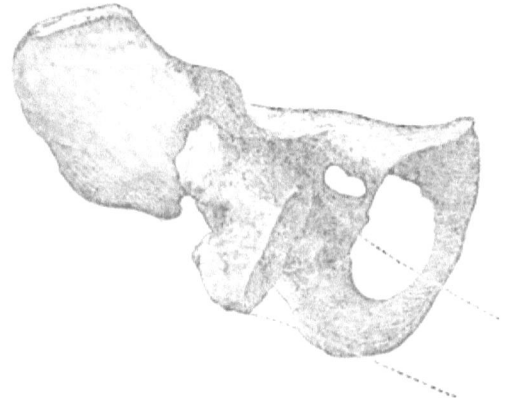

(Fig. 15.)

my curiosity in a pathological point of view. It is astonishing to me how little destruction has been effected by the late suppuration. All the adhesions have been of course carried off, and the bony surfaces in contact with each other are osteoporotic, which is probably the normal condition in connection with the formation of fibrous adhesions. The caput femoris is of course destroyed by the original disease, but the neck has suffered no changes by suppuration since its articulating surface accurately fits in the socket.

I have mentioned these two cases for the purpose of showing that brisement forcé of the anchylosed hip joint is a proceeding not altogether devoid of danger. Nevertheless it is a legitimate operation if performed with due precaution, but the most brilliant results cannot compare with those attainable at the knee and elbow joints.

The previous division of the contracted muscles is to be insisted upon. Myotomy is not only harmless and indispensable to a satisfactory result, it lends also protection against the recurrence of the previous morbus coxarius; and I feel persuaded that Buchring would have had better chances to save his patient had he not omitted that initiatory operation. A few days after the operation we may then proceed to loosen the joint. The patient is to be placed upon the table in the recumbent posture, and when under the full influence of chloroform his pelvis is held by an assistant grasping both sides, with the thumbs upon the anterior superior pinous process of the ilium, whilst the operator presses firmly his foot against the corresponding tuber ischii. Thus prepared, he takes hold of the affected extremity, and with a firm, steady, but gentle traction, extends and abducts the limb. Gentle motions and rotations may be combined with the traction, but they should never be made so powerful or free as to destroy the existing adhesions. We ought to be contented with a good position of the extremity, and not to risk the lives of our patient for the sake of more or less free motion.

In adults there is less danger of recurring disease, and their limbs bear a freer handling.

The fixing of the pelvis is obviously very important to the ulterior results, and the hands of an assistant fail particularly then to fix the pelvis when the thigh is considerably flexed upon the former, for this and the purposes of after treatment, a special apparatus is needed.

Buchring, and subsequently B. Langenbeck, have constructed such apparatus, but they are costly, complicated, cumbersome and inefficient. After various changes and improvements I have succeeded in constructing an apparatus which meets all the requirements, besides being cheap and simple, and may be attached to a plain camp bedstead. The apparatus which I submit to your inspection is much more costly than is necessary (Fig. 16). The essential part of the contrivance is a wooden block accurately adapted to the posterior half of the pelvis, inclusive of the tuber ischii. Any wood carver can make it if you furnish him a plaster of Paris cast. This block is simply lined with chamois, and, if well adapted, the patient can lie in the same for months with the same convenience and ease with which a gum plate with artificial teeth may be worn. When the patient is placed in this block he is fastened down by stout leatherstraps and buckles, in front and across the pelvis. This block is fixed to a plate of sheet iron by means of screws from below; and the iron plate, by means of four bolts, to the frame of the bedstead. Thus you have a simple and complete fixture of the pelvis which lies closely upon the mattress. (Fig. 17.) All that remains is an iron frame at the foot of the bedstead, and two pulleys to shift upon the frame.

(Fig. 16.)

This apparatus should be in readiness when proceeding with *brisement forcé*, and if need be, may at once be used in place of the table and in preference to the manual-fixing of the pelvis.

If you should not succeed in completely extending and abducting the extremity, you may defer the completion and in the meantime keep the limb in the same position in which your first attempt left it, by pulley and weight, or if you have completely succeeded, the after-treatment may at once be fairly commenced. In these cases extension comes in for its profitable employment. Without myotomy and brisement forcé it is more than worthless because dangerous; in combination with those pre-

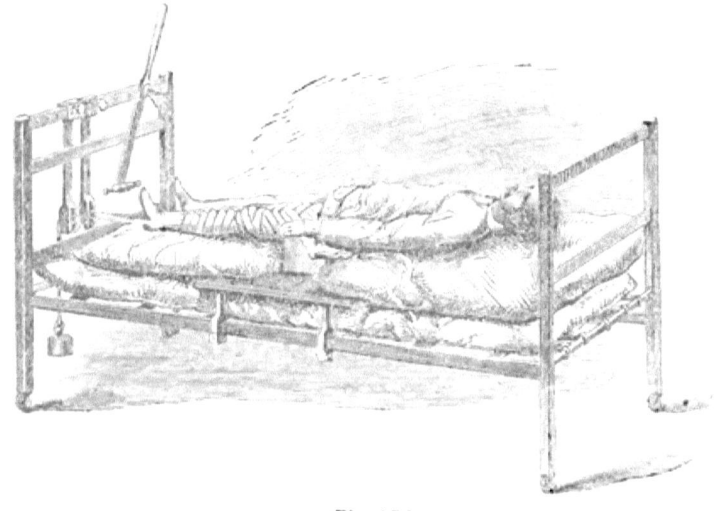

(Fig. 17.)

liminaries it is a most useful auxiliary. Extension with the aid of my apparatus is certainly most efficient and powerful, since the pelvis is completely fixed, and the patient totally prevented from assuming an accomodating position.

I have used it with great benefit in a large number of cases, and know no better substitute.

Two or three months will suffice to render the newly acquired position stable; then you may allow locomotion with the assistance of my portative hip apparatus, with or without crutches as required.

The true bony anchylosis of the hip joint finds its relief in Rhea Barton's operation. I have never had occasion to perform it, and can therefore offer no suggestions drawn from personal experience, but it would seem to me that the attempt at establishing an artificial joint at the line of division is unwarrantable for two reasons: 1st. An artificial joint could never give a sufficient support to the superstructure of the body. 2nd. It inevitably protracts the suppuration with its impending danger of pyæmia.

Sayre, a few years ago, performed this operation, as he alleged with success, but his patient nevertheless died a few months after from pyæmia.

The specimen derived from the case, did not sustain the assertion of that gentleman, no cartilaginous covering, synovial lining or a new capsular ligament having been formed.

Now, gentlemen, I have arrived at the end of our discourse and will finish with relating a few interesting cases. Some of them present peculiar and exceptional clinical features, others may serve as types of their class. Your attention has been most gratifying to me and I feel sincerely thankful for your magnanimous indulgence.

Case I.

Hygroma bursale traumaticum, of eight years standing, fibrous anchylosis of left knee joint with flexed and inverted malposition. (Vide fig. 18 and fig. 12.)

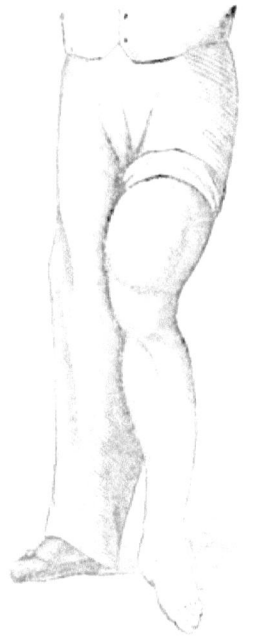

(Fig. 18.)

A young man (Packner) twenty two years old, solicited my professional services in the following case: When at the age of 11 years he sustained an injury to the left knee, which gave him trouble for three years, not materially impeding, however, his locomotion. His general health having materially suffered, his father, a sea-captain, was advised to take the patient on a voyage and give him the benefit of sea-air. On ship board he repeatedly met with falls and slight accidents without being aggravated. One day whilst driving a nail into a plank, the hammer struck him heavily just above the left knee-joint and caused a painful bruise, soon followed by intense agony and swelling.

From that time to the period when I took charge of his case, the patient had never been free from pain and uneasiness, and his haggard, anxious, and desponding appearance bore the unmistakable evidence of severe and continuous suffering. The affected articulation was so tender as to be utterly useless for locomotion; in fact he would not even stand upon the extremity with a mere fraction of the bodily weight put upon it. Hence crutches were requisite, between which the extremity was suspended.

The wealthy father had of course successively consulted the best surgeons he could find, both in Europe and on this continent. They had all agreed in their counsel that amputation was the only remedy.

On examining the affected extremity the following clinical points were elicited:

1. An ovally shaped, smooth and throughout, hard tumor, "9 × 4," inches located immediately above the knee-joint. Its base was broad, abrupt and immovable. There was no tenderness or discoloration about the tumor;

2. The joint was anchylosed but allowed a trifle of motion, which was, however, very painful at its inner circumference;

3. The quadriceps muscle of the thigh was displaced to the outside of the tumor; the patella lodged upon and adhered to the external condyle of the femur;

4. The tibia occupied an angle of 150° with the femur, and was so turned on its axis as to evert the toes;

5. Besides there was a slight inflexion at the knee between the two bones which gave it a knock-kneed appearance;

6. The biceps muscle was considerably shortened and therefore very tense;

7. The temperature of the knee-joint, more particularly at its inner aspect, was not much raised;

8. In fine the affected extremity was moderately attenuated.

The tumor was obviously accountable for the existing articular trouble and malposition. It had raised up and gradually displaced the extensor muscles of the leg. The latter derived additional physical power from acting, as it were, around a pulley, being converted into a flexor, rotator and adductor muscle of the knee. The tibia had yielded to the abnormal traction. The torsion of the joint had set up inflammation of the synovial lining, eventuating in fibrous interarticular adhesion of the articular faces. Reflex contraction of the biceps muscle had ensued. Thus, by the succession of mechanical effects, a most complicated morbid condition had been brought about in course of time, traceable to no other cause than the tumor. The still existing inflammatory action at the inner circumference of the knee-joint, may be ascribed to the abnormal position of the extremity, being diagonal through the femur and bearing the weight of the body upon the internal ligaments.

But the all important diagnostic question centered itself upon the nature of the tumor! The apparently very hard texture suggested bony structure. For ostoid, the tumor was too hard and smooth, and had existed far too long a time to sustain the suspicion of a malignant growth. Periostitis would have circumvented the femur, and not exhibit a broad and flat base. Bone abscess would have distended the femoral tube in all directions and at that size would have become softened. The hardness and smoothness of its surface precluded the idea of an osteosarcoma.

The evidently traumatic cause, the gradual increase, the regular form of the tumor, and the anatomical region, pointed directly and conjointly to the distension of the subcrural bursa. Yet there was no fluctuation, and that ominous hardness was left unaccounted for. Nothwithstanding the discrepancy, I commenced most carefully to explore my ground with the hope of detecting fluctuation; for the rather indefinite supposition suggested itself, that *the resistance of the vagina femoris* might render the tumor both hard and obscure its fluctuation.

At the inner and lower aspect of the growth, a branch of the saphena magna perforated the aponeurosis and dipped into the depth. There I felt some elasticity and very indistinct fluctuation, sufficient evidence of fluid, at any rate, to warrant explorative puncture. The patient, a very intelligent young man, having realized the probable character of his case, and deriving new hope from the proposed proceeding, readily consented to the exploration.

After having made the necessary preparation, I proceeded next day, with some professional friends, to the patient's dwelling. I met with but little encouragement for the operation, either on the part of colleagues or the relatives of the patient. The former dissented *in toto* from the suggestive diagnosis, and the latter presented the authority of the best surgeons of the country as objection to any other proceeding short of amputation of the thigh.

The trocar being inserted, about ℥ xiv of a straw-coloured and alkaline fluid was withdrawn, whereupon the tumor collapsed. On careful examination, the empty sac and its contours could still be discerned; but, of course, the previous hardness had entirely vanished.

Having thus verified the diagnosis, I proceeded with the second part of the programme, *in dividing the outer hamstring, breaking up all articular adhesions, and in fully extending* the extremity. A few minutes served to change the condition of the patient, and infuse him and his friends with new hopes for the future. It could hardly be anticipated that pressure alone would suffice to prevent the re-accumulation of the bursal fluid. In order to close up the old depot, I was induced to inject tincture of iodine.

That operation was followed with violent inflammation and suppuration of the bursa. When, at last, the cavity had closed, the quadriceps muscle was so firmly aglutinated to the thigh-bone, that it seemed indifferent whether the articulation of the knee-joint was re-established or not. The patient, desirous for active life, declared himself quite contented with a straight, useful, and painless, though inflexible extremity, with which he is now able, according to a recent letter to a friend, to walk his forty miles a day, by peddling in California.

The presented photograph fig. 12 is the appearance of the patient at his discharge. At that time I supported his extremity with a straight apparatus, with which the patient now dispenses.

That the hardness of the tumor was simply caused by the constraint and resistance of the vagina femoris, will be admitted without further dispute. And we noticed the *same symptom* in the case of Mr. A., one of the great hotel proprietors of New York. We need hardly say that the correct treatment of Mr. A.'s case depended likewise on correct discernment of the tumor, about whose character and structure conflicting opinions and apprehensions had been advanced.

CASE II.

Traumatic diastasis of the lower epiphysis of left femur. Remarkable deformity and malposition of the knee-joints. Abnormal lateral mobility. Total resection. Recovery.

Francis Shaw, a lad of fourteen years, of Irish descent, and endowed with robust health, presented himself in October 1860, at the clinic of the Brooklyn Medical and Surgical Institute. He came at the instigation of a surgical instrument maker to get my advice with reference to the feasibility of a mechanical apparatus to steady and support his limb, and to render it useful for locomotion. He stated that he had acquired the deformity when but seven years old, and that ever since the trouble had increased, and that then he was unable to use his extremity to any purpose. To the best of his memory he received a blow at the knee-joint with an iron rod, which gave him pain and disabled him for a short time. A physician had been called in soon after the injury, but finding no undue mobility or deformity he pronounced it a simple contusion, and advised rest and cold fomentations. These directions were followed for three weeks, when the patient resumed his walk.

Since that time dates the impediment. In the erect posture, the patient throws his whole weight upon the sound member, when balanced between two chairs a three inch block is required to equalise the length of both extremities, as may be seen in the adjoining diagram (Fig. 19). The left limb is peculiarly knock-kneed, the thigh being adducted, the leg abducted and everted, and laterally both forming an angle of 120°. This position alone would have been quite sufficient to render locomotion infirm and defective, but as it was, the limb became totally useless by the relaxation of the knee joint. At the moment the patient rested upon the affected extremity, the leg became still more abducted and everted, and the angle with the thigh could easily be reduced to 80° and less. Both

(Fig. 19. See page 299.)

articular faces moved with undue freedom over each other, and the tibia could be freely rotated upon the femur, the scope of eversion being, however, greater. This abnormal condition was due to some remarkable anatomical changes in the configuration of the joint. The articular surface of the femur had an oblique direction, from below and inward to up and outward, the two condyles were absent, and the bone terminated below as a segment of a sphere, of which but a part was appropriated for articulating purposes, the patella and the quadriceps muscle were drawn out of position towards the outer aspect of the extremity. The tendon of the biceps muscle occupied the popliteal space. In every other respect the limb presented the ordinary condition, except being slightly attenuated.

Before the patient had applied to our institution he had presented himself before the surgical staff of the New-York City Hospital, who had come to the conclusion to advise mechanical support, which was, however, entirely out of the question. On the other hand Francis Shaw had arrived at an age which made him desirous of entering upon some business, and therefore insisted upon some means to render his limb serviceable. There was nothing left but the exsection of the knee joint or amputation of the thigh; for no orthopædic treatment could be relied upon to materially alter the anatomical status.

I could not hesitate to decide in favour of exsection, since both the constitution of the lad as well as the bony structure concerned, were in a most auspicious condition. The operation was performed on the 9th of October. I had to remove quite a large piece from the femur so as to obtain a rectangular surface; but a very thin slice was taken from the tibia, the patella was likewise removed. The bones were then brought in close proximity and kept in position by softened iron wire, and the wound united by silver wire in fine, the limb was secured in one of the iron splints (vide fig. 10) which left the knee-joint itself free of access. Recovery followed rapidly, partly by first intention. The bone wire was removed on the twenty sixth day after the operation, and at the end of the second month the patient was up and about, and accompanied me on crutches to a neighbouring gallery to have his photograph taken. Represented in (Fig. 20.)

(Fig. 20.)

On the 28th Feb. 1861, I. exhibited Francis Shaw at the New-York Pathological society, when his conditions were as follows: integuments, completely cicatrized; firm union of the bones by short fibrous tissues admitting but of scanty motion; moderate enlargement of the circumference; circulation and temperature normal; deficiency in length two inches; correct position of the foot. With a heel of two and a quarter inches, pelvis and shoulders stand square. His locomotion was, aside from the stiffness of his knee, unimpeded.

You may imagine that the diagnosis of the case must have been perplexing, when the most distinguished surgeons of New-York signally failed to realize it, nor could I lay any claim to a clear understanding of the proximate cause in the premises before the operation, yet I have the gratification to say that the views I had first formed and expressed to my class, did not fall short of the reality.

That the injury to Francis Shaw had produced no fracture was self-evident from the previous history so clearly related. Nevertheless the continuity of the femur must have suffered in such a manner as not to disturb the form of the limb, nor give rise to any undue mobility. With diastasis of the lower femoral epiphysis these conditions are compatible. Had the patient quietly remained in bed for six or eight weeks, there is no doubt that the subsequent trouble would have been averted. But in rising prematurely, the soft agglutination of the epiphysis with the shaft gave way and allowed the former to turn gradually round, and with it dislodge the entire joint. In the newly acquired position the undue pressure upon the external condyle of the femur had gradually diminished its size until no trace was left. And the internal condyle became the terminating end of the femur. The fragments of bone removed by the operation (fig. 21 and 22*), render this reasoning at least plausible if not conclusive.

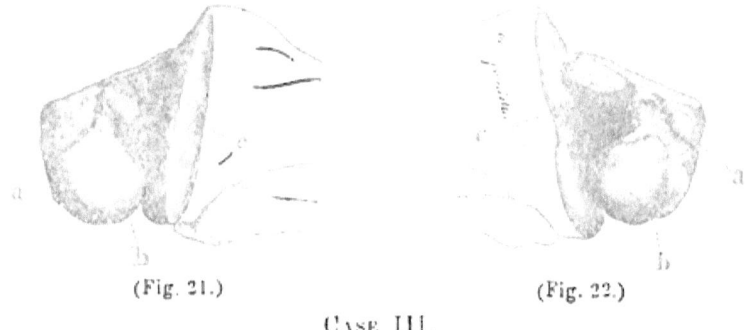

(Fig. 21.)   (Fig. 22.)

CASE III.

Morbus Coxarius in its third stage. Consecutive Abscess connecting with the joint.—Complete prevention of malposition.

George D., ten years old, of good constitution and general health, descending from healthy parents, and one of nine children who are enjoying perfect health, came twenty months ago under my treatment. His

---

* *a*. Epiphyseal line.
   b. Internal condyle of femur.
   c. Slice of tibia.
   Fig. 22, represents the posterior view.

left hip-joint was then very tender and immovable, the extremity was slightly bent upon the pelvis, abducted and rotated with eversion of the toes. The pelvis was lowered at the affected side, and the spine consequently inclined the same way. On examination under chloroform, it was found that the hip-joint was almost immovable, allowing but slight flexion and extension, but no adduction and rotation whatsoever. The tensor vaginæ femoris and the pectinæus muscle contracted.

There was but a moderate fluctuation at the joint. In addition to this I was informed that the patient complained of pain at the knee and violent nocturnal paroxysms. The limb was moderately attenuated. Although the boy had manifested the symptoms of morbus coxarius but a very short time, he gave evidence of constitutional suffering, looked pale and thin. A fall directly upon the left hip, was assigned as the ostensible cause of this disease.

These symptoms strictly coincided with the second stage of morbus coxarius.

The treatment was initiated with leeches to the affected articulation. The contracted muscles were thereupon divided and the patient was placed in the wire apparatus, and thus rest and position of the extremity insured.

The immediate effect of this treatment manifested itself in complete repose and immunity from pain, both structural and reflected. This treatment was continued for six months, when again a thorough examination was instituted. There was almost complete mobility, without crepitus; no fluctuation about the joint; the limb occupies a rectangular position to the pelvis. There was no pain on pressure or motion. The constitutional appearance of the patient was notably improved, appetite and rest were perfect.

Presuming that the disease had been effectually arrested, I allowed the patient one hour's locomotion per day, with the hip splint and crutches, and this time to be gradually prolonged provided no active symptoms should recur. During the balance of the day and the night, in the recumbent posture, and the limb again secured as before. There was no reason to alter the plan, and at the end of another six months he enjoyed his full freedom and went regularly to school, crutches and portative apparatus, as well as the wire apparatus during the night, being continued.

About four months ago, an abscess formed over the place where the tensor vaginæ femoris had been divided, and was attended with the ordinary signs. It was punctured, evacuated, and its walls kept compressed by flannel bandage; since then it has three times refilled and again been punctured. Each time the wound closed. The matter drawn from the abscess was rather thin and somewhat soapy, containing, however, no structural detritus of any account, and particularly no elements of bone. I am rather undecided as to the nature and meaning of the abscess, and have no means of ascertaining whether it connects with the joint or is the consequence of suppurative bursitis. There is indeed not a single symptom indicative of the joint being implicated, although the possibility cannot be denied. But the fact that the punctures close and form no sinuses, is rather against articular suppuration. It is at best therefore an open question.

On the other hand I have seen these abscesses often form at the same location, and subsequent to the division of the tensor vaginæ femoris. Not unlikely these abscesses grow out of an injury to the bursa of that muscle, and would have no great pathological import. If this version should prove true, the diagnosis of this case should be modified accordingly. From the general aspect of the case, I expect perfect recovery at no distant time. The diagrams (figs. 23—24) represent the present

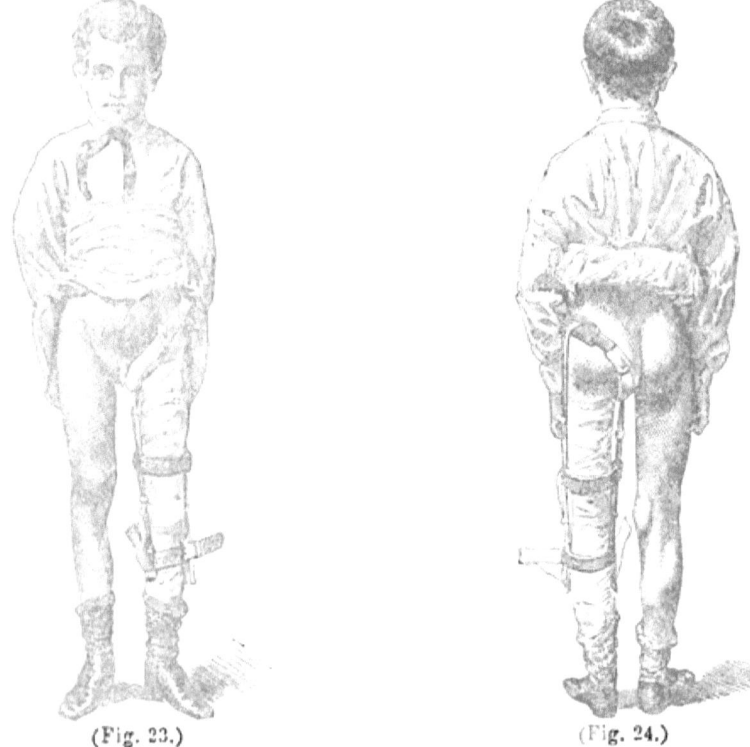

(Fig. 23.) (Fig. 24.)

status of my patient in as far as the position of the affected limb is concerned, and it will be observed that form, position, and length are normal, not even the circumference of the limb differs materially with its fellow.

Case IV.

Malposition of the right limb with more than four inches shortening, the result of now extinct Hip Disease.

Harry M., eleven years of age, came under my charge in the following condition. The right extremity considerably attenuated; the thigh without its proper contours; extreme adduction and inversion; pelvis tilted up and rotated backward; corresponding deflection of the spine; gait very awkward and limping, in spite of a four inch heeled boot; trochanter major protrudes considerably, and exceeds by three quarters of an inch a line drawn between the anterior superior spinous process of the ilium and the tuber ischii; insignificant mobility of the articulation, without a trace of abduction and rotation.

These impediments were the consequences of morbus coxarius, since eighteen months entirely extinct.

Although of slender build, he had enjoyed perfect health, and been a very active boy up to the very time when he was suddenly struck down with that disease.

There is no morbid diathesis in the family; the father of the patient is even a very robust, muscular and active man, the very picture of health and manliness. In addition to this the patient has been, and is still, surrounded with the attributes of opulence and rational hygiene. The premonitory symptoms were but few, insignificant and of short duration. When at a boarding school in the country the patient was suddenly attacked with the most violent symptoms of morbus coxarius, which continued with unabated intensity for five months; then they almost as abruptly abated, leaving the patient in that deformed state which I have briefly sketched. But the shortening had steadily increased so as to require from time to time a higher heel to his boot. Even during the 6 months preceding the operation, the increasing shortening of the limb had been observed. He had, however, completely regained his standard of excellent constitutional health, and was as active as before. There were no local symptoms indicative of continued joint disease.

I have not been able to ascertain the cause of the original attack. There is certainly no pretence of constitutional causation, although the patient does not remember having met with any accident worth speaking of. I, nevertheless, consider myself justified in assuming the same, for the very activity of the patient seems to warrant such a supposition, still more so the violent character of the disease and its rapid course without suppuration.

The patient came under my treatment in the spring of this year, and remained four months with me. During this time I have divided successively most of the abductor muscles; and at four different occasions, with the assistance of chloroform, broken down most fibrous adhesions, and by steady extension in the recumbent posture and repeated passive motions, I have succeeded in placing the affected extremity in a rectangular position to the pelvis, and extended and loosened the still existing fibrous impediments to such a degree as to allow moderate mobility of the articulation.

From the high position and prominence of the larger trochanter, it is evident, that the neck of the femur rides upon and is fastened to a new articular facet at the superior and posterior portion of the acetabular margin, where it still remains, and from which position I do not intend to displace it. At the end of the second month I allowed locomotion to the patient, supported by crutches and my first hip apparatus. It was at that time that the photograph (figs. 24 and 5) were taken. You may judge for

(Fig. 25.)

yourselves of the material changes towards improvement which had been effected up to this time. Previous to his discharge, another photograph with the second hip instrument applied, was obtained, (vide fig. 7). In that position the pelvis has resumed its proper level, the extremity stands rect-angularly, within five eighths of an inch off the floor. The passive motions are still continued with due care, and daily lubrications are being made with phosphorated oil, to promote healthful innervation and nutrition.

The patient is directed to wear the hip instrument night and day until the changes of form and position become permanent, when a heel $\frac{5}{8}$ of an inch higher than that of the other boot, will suffice to ensure easy gait and locomotion.

These changes have been wrought within the short period of four months in a deformity and malposition which in former times were considered beyond surgical aid, and this case furnishes, therefore, an illustration of the grand progress in orthopædic surgery.

Brooklyn, N.Y., Clinton, corner of Warren steeet.

www.ingramcontent.com/pod-product-compliance
Lightning Source LLC
Chambersburg PA
CBHW020154170426
43199CB00010B/1032